THE ROCKY COAST

Books by Rachel Carson

THE SEA AROUND US

UNDER THE SEA WIND

THE EDGE OF THE SEA

SILENT SPRING

THE SENSE OF WONDER

THE ROCKY COAST

by Rachel Carson

photographs by Charles Pratt

drawings by Bob Hines

The McCall Publishing Company

New York

To Dorothy and Stanley Freeman
*who have gone down with me into the low-tide world
and have felt its beauty and its mystery.*

—Dedication from the Original Edition

ACKNOWLEDGMENTS

From the Original Edition

Our understanding of the nature of the shore and of the lives of sea animals has been acquired through the labor of many hundreds of people, some of whom have devoted a lifetime to the study of a single group of animals. In my researches for this book I have been deeply conscious of the debt of gratitude we owe these men and women, whose toil allows us to sense the wholeness of life as it is lived by many of the creatures of the shore. I am even more immediately aware of my debt to those I have consulted personally, comparing observations, seeking advice and information and always finding it freely and generously given. It is impossible to express my thanks to all these people by name, but a few must have special mention. Several members of the staff of the United States National Museum have not only settled many of my questions but have given invaluable advice and assistance to Bob Hines in his preparation of the drawings. For this help we are especially grateful to R. Tucker Abbott, Frederick M. Bayer, Fenner Chace, the late Austin H. Clark, Harald Rehder, and Leonard Schultz. Dr. W. N. Bradley of the United States Geological Survey has been my friendly advisor on geological matters, answering many questions and critically reading portions of the manuscript. Professor William Randolph Taylor of the University of Michigan has responded instantly and cheerfully to my calls for aid in identifying marine algae, and Professor and Mrs. T. A. Stephenson of the University College of Wales, whose work on the ecology of the shore

vii

has been especially stimulating, have advised and encouraged me in correspondence. To Professor Henry B. Bigelow of Harvard University I am everlastingly in debt for encouragement and friendly counsel over many years. The grant of a Guggenheim Fellowship helped finance the first year of study in which the foundations of this book were laid, and some of the field work that has taken me along the tide lines from Maine to Florida.

PREFACE

From the Original Edition

Like the sea itself, the shore fascinates us who return to it, the place of our dim ancestral beginnings. In the recurrent rhythms of tides and surf and in the varied life of the tide lines there is the obvious attraction of movement and change and beauty. There is also, I am convinced, a deeper fascination born of inner meaning and significance.

When we go down to the low-tide line, we enter a world that is as old as the earth itself—the primeval meeting place of the elements of earth and water, a place of compromise and conflict and eternal change. For us as living creatures it has special meaning as an area in or near which some entity that could be distinguished as Life first drifted in shallow waters— reproducing, evolving, yielding that endlessly varied stream of living things that has surged through time and space to occupy the earth.

To understand the shore, it is not enough to catalogue its life. Understanding comes only when, standing on a beach, we can sense the long rhythms of earth and sea that sculptured its land forms and produced the rock and sand of which it is composed; when we can sense with the eye and ear of the mind the surge of life beating always at its shores— blindly, inexorably pressing for a foothold. To understand the life of the shore, it is not enough to pick up an empty shell and say "This is a murex," or "That is an angel wing." True understanding demands intuitive comprehension of the whole life of the creature that once inhabited this empty shell: how it survived amid surf and storms, what were its enemies, how it found food and reproduced its kind, what were its relations to the particular sea world in which it lived.

The seashores of the world may be divided into three basic types: the rugged shores of rock, the sand beaches, and the coral reefs and all their associated features. Each has its typical community of plants and animals. The Atlantic coast of the United States is one of the few in the world that provide clear examples of each of these types. I have chosen it as the setting for my pictures of shore life, although—such is the universality of the sea world—the broad outlines of the pictures might apply on many shores of the earth.

A Note from the Photographer

For about fifteen summers as an adult and six as a child, I have hung around the intertidal area of the Maine coast. I wanted to celebrate what I saw, and the resulting photographs I took to Rachel Carson. Apologizing for her lack of critical vocabulary in photography, she said about them that "this is the way it really looks." I thought she said it with surprise that a picture could touch the actuality of her own experience rather than just be admired on aesthetic grounds. It remains the greatest compliment my pictures have ever received. If my experience of a tidal pool really met hers I must have got to know that particular tidal pool very well—which, to me anyway, is what photography is all about. By then I'd learned that it is only by the closest attention to what is in front of one—rather than what is "inside" of one—that a photograph can be made, and I'd gone a certain distance in learning how to give this attention to the objective world.

I think that Rachel Carson had discovered where she wanted to spend most of her time, and everything else followed: scientific exploration, exact description, and finally *Silent Spring*. Many people start a career with passion, but few end it with the same passion intact, much less retain the center of the passion throughout.

—*Charles Pratt*

THE ROCKY COAST

When the tide is high on a rocky shore, when its brimming fullness creeps up almost to the bayberry and the junipers where they come down from the land, one might easily suppose that nothing at all lived in or on or under these waters of the sea's edge. For nothing is visible. Nothing except here and there a little group of herring gulls, for at high tide the gulls rest on ledges of rock, dry above the surf and the spray, and they tuck their yellow bills under their feathers and doze away the hours of the rising water. Then all the creatures of the tidal rocks are hidden from view, but the gulls know what is there, and they know that in time the water will fall away again and give them entrance to the strip between the tide lines.

When the tide is rising the shore is a place of unrest, with the surge leaping high over jutting rocks and running in lacy cascades of foam over the landward side of massive boulders. But on the ebb it is more peaceful, for then the waves do not have behind them the push of the inward pressing tides. There is no particular drama about the turn of the tide, but presently a zone of wetness shows on the gray rock slopes, and offshore the incoming swells begin to swirl and break over hidden ledges. Soon the rocks that the high tide had concealed rise into view and glisten with the wetness left on them by the receding water.

Small, dingy snails move about over rocks that are slippery with the growth of infinitesimal green plants; the snails scraping, scraping, scraping to find food before the surf returns.

Like drifts of old snow no longer white, the barnacles come into view; they blanket rocks and old spars wedged into rock crevices, and their sharp cones are sprinkled over empty mussel shells and lobster-pot buoys and the hard stipes of deep-water seaweeds, all mingled in the flotsam of the tide.

Meadows of brown rockweeds appear on the gently sloping rocks of the shore as the tide imperceptibly ebbs. Smaller patches of green weed, stringy as mermaids' hair, begin to turn white and crinkly where the sun has dried them.

Now the gulls, that lately rested on the higher ledges, pace with grave intentness along the walls of rock, and they probe under the hanging curtains of weed to find crabs and sea urchins.

In the low places little pools and gutters are left where the water trickles and gurgles and cascades in miniature waterfalls, and many of the dark caverns between and under the rocks are floored with still mirrors holding the reflections of delicate creatures that shun the light and avoid the shock of waves—the cream-colored flowers of the small anemones and the pink fingers of soft coral, pendent from the rocky ceiling.

In the calm world of the deeper rock pools, now undisturbed by the tumult of incoming waves, crabs sidle along the walls, their claws busily touching, feeling, exploring for bits of food. The pools are gardens of color composed of the delicate green and ocher-yellow of encrusting sponge, the pale pink of hydroids that stand like clusters of fragile spring flowers, the bronze and electric-blue gleams of the Irish moss, the old-rose beauty of the coralline algae.

And over it all there is the smell of low tide, compounded of the faint, pervasive smell of worms and snails and jellyfish and crabs—the sulphur smell of sponge, the iodine smell of rockweed, and the salt smell of the rime that glitters on the sun-dried rocks.

One of my own favorite approaches to a rocky seacoast is by a rough path through an evergreen forest that has its own peculiar enchantment. It is usually an early morning tide that takes me along that forest path, so that

5

Sandpiper at low tide

the light is still pale and fog drifts in from the sea beyond. It is almost a ghost forest, for among the living spruce and balsam are many dead trees—some still erect, some sagging earthward, some lying on the floor of the forest. All the trees, the living and the dead, are clothed with green and silver crusts of lichens. Tufts of the bearded lichen or old man's beard hang from the branches like bits of sea mist tangled there. Green woodland mosses and a yielding carpet of reindeer moss cover the ground. In the quiet of that place even the voice of the surf is reduced to a whispered echo and the sounds of the forest are but the ghosts of sound—the faint sighing of evergreen needles in the moving air; the creaks and heavier groans of half-fallen trees resting against their neighbors and rubbing bark against bark; the light rattling fall of a dead branch broken under the feet of a squirrel and sent bouncing and ricocheting earthward.

But finally the path emerges from the dimness of the deeper forest and comes to a place where the sound of surf rises above the forest sounds—the hollow boom of the sea, rhythmic and insistent, striking against the rocks, falling away, rising again.

Up and down the coast the line of the forest is drawn sharp and clean on the edge of a seascape of surf and sky and rocks. The softness of sea fog blurs the contours of the rocks; gray water and gray mists merge offshore in a dim and vaporous world that might be a world of creation, stirring with new life.

The sense of newness is more than illusion born of the early morning light and the fog, for this is in very fact a young coast. It was only yesterday in the life of the earth that the sea came in as the coast subsided, filling the valleys and rising about the slopes of the hills, creating these rugged shores where rocks rise out of the sea and evergreen forests come down to the coastal rocks. Once this shore was like the ancient land to the south, where the nature of the coast has changed little during the millions of years since the sea and the wind and the rain created its sands and shaped them into dune and beach and offshore bar and shoal. The northern coast, too, had its flat coastal plain bordered by wide beaches of sand. Behind these lay a landscape of rocky hills alternating with valleys

Bearded lichen

that had been worn by streams and deepened and sculptured by glaciers. The hills were formed of gneiss and other crystalline rocks resistant to erosion; the lowlands had been created in beds of weaker rocks like sandstones, shale, and marl.

Then the scene changed. From a point somewhere in the vicinity of Long Island the flexible crust of the earth tilted downward under the burden of a vast glacier. The regions we know as eastern Maine and Nova Scotia were pressed down into the earth, some areas being carried as much as 1200 feet beneath the sea. All of the northern coastal plain was drowned. Some of its more elevated parts are now offshore shoals, the fishing banks off the New England and Canadian coasts—Georges, Browns, Quereau, the Grand Bank. None of it remains above the sea except here and there a high and isolated hill, like the present island of Monhegan, which in ancient times must have stood above the coastal plain as a bold monadnock.

Where the mountainous ridges and the valleys lay at an angle to the coast, the sea ran far up between the hills and occupied the valleys. This was the origin of the deeply indented and exceedingly irregular coast that is characteristic of much of Maine. The long narrow estuaries of the Kennebec, the Sheepscot, the Damariscotta and many other rivers run inland a score of miles. These salt-water rivers, now arms of the sea, are the drowned valleys in which grass and trees grew in a geologic yesterday. The rocky, forested ridges between them probably looked much as they do today. Offshore, chains of islands jut out obliquely into the sea, one beyond another—half-submerged ridges of the former land mass.

But where the shore line is parallel to the massive ridges of rock the coast line is smoother, with few indentations. The rains of earlier centuries cut only short valleys into the flanks of the granite hills, and so when the sea rose there were created only a few short, broad bays instead of long winding ones. Such a coast occurs typically in southern Nova Scotia, and also may be seen in the Cape Ann region of Massachusetts, where the belts of resistant rock curve eastward along the coast. On such a coast, islands, where they occur, lie parallel to the shore line instead of putting boldly out to sea.

As geologic events are reckoned, all this happened rather rapidly and suddenly, with no time for gradual adjustment of the landscape; also it happened quite recently, the present relation of land and sea being achieved perhaps no more than ten thousand years ago. In the chronology of Earth, a few thousand years are as nothing, and in so brief a time the waves have prevailed little against the hard rocks that the great ice sheet scraped clean of loose rock and ancient soil, and so have scarcely marked out the deep notches that in time they will cut in the cliffs.

For the most part, the ruggedness of this coast is the ruggedness of the hills themselves. There are none of the wave-cut stacks and arches that distinguish older coasts or coasts of softer rock. In a few, exceptional places the work of the waves may be seen. The south shore of Mount Desert Island is exposed to heavy pounding by surf; there the waves have cut out Anemone Cave and are working at Thunder Hole to batter through the roof of the small cave into which the surf roars at high tide.

In places the sea washes the foot of a steep cliff produced by the shearing effect of earth pressure along fault lines. Cliffs on Mount Desert—Schooner Head, Great Head, and Otter—tower a hundred feet or more above the sea. Such imposing structures might be taken for wave-cut cliffs if one did not know the geologic history of the region.

On the coasts of Cape Breton Island and New Brunswick the situation is very different and examples of advanced marine erosion occur on every hand. Here the sea is in contact with weak rock lowlands formed in the Carboniferous period. These shores have little resistance to the erosive power of the waves, and the soft sandstone and conglomerate rocks are being cut back at an annual rate averaging five or six inches, or in some places several feet. Marine stacks, caves, chimneys, and archways are common features of these shores.

Here and there on the predominantly rocky coast of northern New England there are small beaches of sand, pebbles, or cobblestones. These have a varied origin. Some came from glacial debris that covered the rocky surface when the land tilted and the sea came in. Boulders and pebbles often are carried in from deeper water offshore by seaweeds that

have gripped them firmly with their "holdfasts." Storm waves then dislodge weed and stone and cast them on the shore. Even without the aid of weeds, waves carry in a considerable volume of sand, gravel, shell fragments, and even boulders. These occasional sandy or pebbly beaches are almost always in protected, incurving shores or dead-end coves, where the waves can deposit debris but from which they cannot easily remove it.

When, on those coastal rocks between the serrate line of spruces and the surf, the morning mists conceal the lighthouses and fishing boats and all other reminders of man, they also blur the sense of time and one might easily imagine that the sea came in only yesterday to create this particular line of coast. Yet the creatures that inhabit the intertidal rocks have had time to establish themselves here, replacing the fauna of the beaches of sand and mud that probably bordered the older coast. Out of the same sea that rose over the northern coast of New England, drowning the coastal plain and coming to rest against the hard uplands, the larvae of the rock dwellers came—the blindly searching larvae that drift in the ocean currents ready to colonize whatever suitable land may lie in their path or to die, if no such landfall is their lot.

Although no one recorded the first colonist or traced the succession of living forms, we may make a fairly confident guess as to the pioneers of the occupation of these rocks, and the forms that followed them. The invading sea must have brought the larvae and young of many kinds of shore animals, but only those able to find food could survive on the new shore. And in the beginning the only available food was the plankton that came in renewed clouds with every tide that washed the coastal rocks. The first permanent inhabitants must have been such plankton-strainers as the barnacles and mussels, who require little but a firm place to which they may attach themselves. Around and among the white cones of the barnacles and the dark shells of the mussels it is probable that the spores of algae settled, so that a living green film began to spread over the upper rocks. Then the grazers could come—the little herds of snails that laboriously scrape the rocks with their sharp tongues, licking off the nearly

invisible covering of tiny plant cells. Only after the establishment of the plankton-strainers and the grazers could the carnivores settle and survive. The predatory dog whelks, the starfish, and many of the crabs and worms must, then, have been comparative latecomers to this rocky shore. But all of them are there now, living out their lives in the horizontal zones created by the tides, or in the little pockets or communities of life established by the need to take shelter from surf, or to find food, or to hide from enemies.

The pattern of life spread before me when I emerge from that forest path is one characteristic of exposed shores. From the edge of the spruce forests down to the dark groves of the kelps, the life of the land grades into the life of the sea, perhaps with less abruptness than one would expect, for by various little interlacing ties the ancient unity of the two is made clear.

Lichens live in the forest above the sea, in the silent intensity of their toil crumbling away the rocks as lichens have done for millions of years. Some leave the forest and advance over the bare rock toward the tide line; a few go even farther, enduring a periodic submersion by the sea so that they may work their strange magic on the rocks of the intertidal zone. In the dampness of foggy mornings the rock tripe on the seaward slopes is like sheets of thin, pliable green leather, but by midday under a drying sun it has become blackened and brittle; then the rocks look as though they were sloughing off a thin outer layer. Thriving in the salt spray, the wall lichen spreads its orange stain on the cliffs and even on the landward side of boulders that are visited by the highest tides of each moon. Scales of other lichens, sage-green, rolled and twisted into strange shapes, rise from the lower rocks; from their under surfaces black, hairy processes work down among the minute particles of rock substance, giving off an acid secretion to dissolve the rock. As the hairs absorb moisture and swell, fine grains of the rock are dislodged and so the work of creating soil from the rock is advanced.

Below the forest's edge the rock is white or gray or buff, according to its mineral nature. It is dry and belongs to the land; except for a few

Wall lichen

insects or other land creatures using it as pathways to the sea it is barren. But just above the area that clearly belongs to the sea, it shows a strange discoloration, being strongly marked with streaks or patches or continuous bands of black. Nothing about this black zone suggests life; one would call it a dark stain, or at most a felty roughening of the rock surface. Yet it is actually a dense growth of minute plants. The species that compose it sometimes include a very small lichen, sometimes one or more of the green algae, but most numerously the simplest and most ancient of all plants, the blue-green algae. Some are enclosed in slimy sheaths that protect them from drying and fit them to endure long exposure to sun and air. All are so minute as to be invisible as individual plants. Their gelatinous sheaths and the fact that the whole area receives the spray of breaking waves make this entrance to the sea world slippery as the smoothest ice.

This black zone of the shore has a meaning above and beyond its drab and lifeless aspect—a meaning obscure, elusive, and infinitely tantalizing. Wherever rocks meet the sea, the microplants have written their dark inscription, a message only partially legible although it seems in some way to be concerned with the universality of tides and oceans. Though other elements of the intertidal world come and go, this darkening stain is omnipresent. The rockweeds, the barnacles, the snails, and the mussels appear and disappear in the intertidal zone according to the changing nature of their world, but the black inscriptions of the microplants are always there. Seeing them here on this Maine coast, I remember how they also blackened the coral rim of Key Largo, and streaked the smooth platform of coquina at St. Augustine, and left their tracings on the concrete jetties at Beaufort. It is the same all over the world—from South Africa to Norway and from the Aleutians to Australia. This is the sign of the meeting of land and sea.

Once below the dark film, I begin to look for the first of the sea creatures pressing up to the threshold of the land. In seams and crevices in the high rocks I find them—the smallest of the periwinkle tribe, the rock or rough periwinkle. Some—the infant snails—are so small that I need my

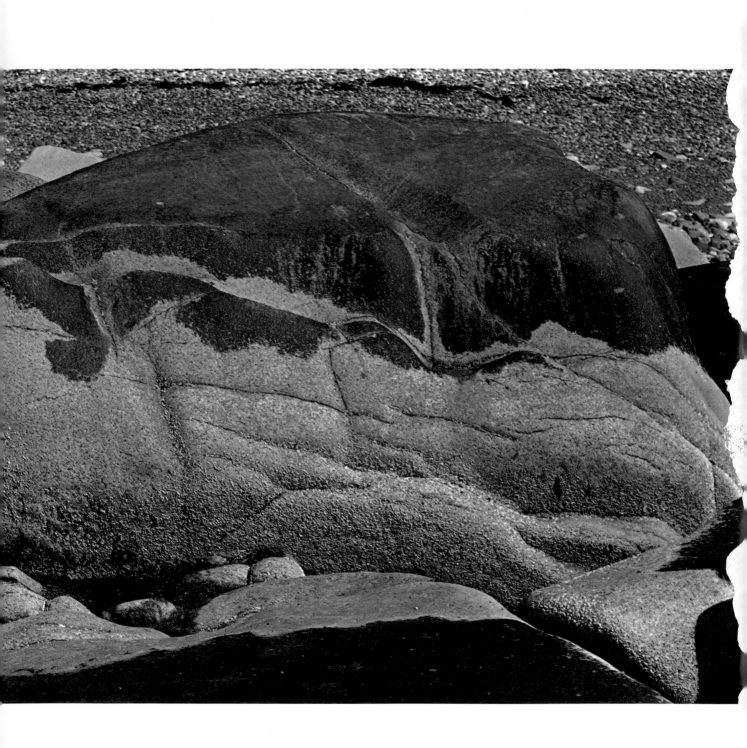

hand lens to see them clearly; and among the hundreds that crowd into these cracks and depressions I can find a gradation of sizes up to the half-inch adults. If these were sea creatures of ordinary habits, I would think the small snails were young produced by some distant colony and drifted here as larvae after spending a period at sea. But the rough periwinkle sends no young into the sea; instead it is a viviparous species and the eggs, each encased within a cocoon, are held within the mother while they develop. The contents of the cocoon nourish the young snail until finally it breaks through the egg capsule and then emerges from the mother's body, a completely shelled little creature about the size of a grain of finely ground coffee. So small an animal might easily be washed out to sea; hence, no doubt, the habit of hiding in crevices and in empty barnacle shells, where often I have found them in numbers.

At the level where most of the rough periwinkles live, however, the sea comes only every fortnight on the spring tides, and in the long intervals the flying spray of breaking waves is their only contact with the water. While the rocks are thoroughly wet with spray the periwinkles can spend much time out on the rocks feeding, often working well up into the black zone. The microplants that create the slippery film on the rocks are their food; like all snails of their group, the periwinkles are vegetarians. They feed by scraping the rocks with a peculiar organ set with many rows of sharp, calcareous teeth. This organ, the radula, is a continuous belt or ribbon that lies on the floor of the pharynx. If unwound, it would be many times the length of the animal, but it is tightly coiled like a watch spring. The radula itself consists of chitin, the substance of insects' wings and lobsters' shells. The teeth that stud it are arranged in several hundred rows (in another species, the common periwinkle, the teeth total about 3500). A certain amount of wear is involved in scraping the rocks, and when the teeth in current use are worn down, an endless supply of new ones can be rolled up from behind.

And there is wear, also, on the rocks. Over the decades and the centuries, a large population of periwinkles scraping the rocks for food has a pronounced erosive effect, cutting away rock surfaces, grain by grain, deepening the tide pools. In a tide pool observed for sixteen years by a

24

California biologist, periwinkles lowered the floor about three-eighths of an inch. Rain, frost, and floods—the earth's major forces of erosion—operate on approximately such a scale.

The periwinkles grazing on the intertidal rocks, waiting for the return of the tide, are poised also in time, waiting for the moment when they can complete their present phase of evolution and move forward onto the land. All snails that are now terrestrial came of marine ancestry, their forebears having at some time made the transitional crossing of the shore. The periwinkles now are in mid-passage. In the structure and habits of the three species found on the New England coast, one can see clearly the evolutionary stages by which a marine creature is transformed into a land dweller. The smooth periwinkle, still bound to the sea, can endure only brief exposure. At low tide it remains in wet seaweeds. The common periwinkle often lives where it is submerged only briefly at high tide. It still sheds eggs into the sea and so is not ready for land life. The rough periwinkle, however, has cut most of the ties that confine it to the sea; it is now almost a land animal. By becoming viviparous it has progressed beyond dependence on the sea for reproduction. It is able to thrive at the level of the high water of the spring tides because, unlike the related periwinkles of lower tidal levels, it possesses a gill cavity that is well supplied with blood vessels and functions almost as a lung to breathe oxygen from the air. Constant submersion is, in fact, fatal to it and at the present stage of its evolution it can endure up to thirty-one days of exposure to dry air.

The rough periwinkle has been found by a French experimenter to have the rhythm of the tides deeply impressed upon its behavior patterns, so that it "remembers" even when no longer exposed to the alternating rise and fall of the water. It is most active during the fortnightly visits of the spring tides to its rocks, but in the waterless intervals it becomes progressively more sluggish and its tissues undergo a certain desiccation. With the return of the spring tides the cycle is reversed. When taken into a laboratory the snails for many months reflect in their behavior the advance and retreat of the sea over their native shores.

On this exposed New England coast the most conspicuous animals of

Barnacles on an exposed shore

the high-tide zone are the rock or acorn barnacles, which are able to live in all but the most tumultuous surf. The rockweeds here are so stunted by wave action that they offer no competition, and so the barnacles have taken over the upper shore, except for such space as the mussels have been able to hold.

At low tide the barnacle-covered rocks seem a mineral landscape carved and sculptured into millions of little sharply pointed cones. There is no movement, no sign or suggestion of life. The stony shells, like those of mollusks, are calcareous and are secreted by the invisible animals within. Each cone-shaped shell consists of six neatly fitted plates forming an encircling ring. A covering door of four plates closes to protect the barnacle from drying when the tide has ebbed, or swings open to allow it to feed. The first ripples of incoming tide bring the petrified fields to life. Then, if one stands ankle-deep in water and observes closely, one sees tiny shadows flickering everywhere over the submerged rocks. Over each individual cone, a feathered plume is regularly thrust out and drawn back within the slightly opened portals of the central door—the rhythmic motions by which the barnacle sweeps in diatoms and other microscopic life of the returning sea.

The creature inside each shell is something like a small pinkish shrimp that lies head downward, firmly cemented to the base of this chamber it cannot leave. Only the appendages are ever exposed—six pairs of branched, slender wands, jointed and set with bristles. Acting together, they form a net of great efficiency.

The barnacle belongs to the group of arthropods known as the Crustacea, a varied horde including the lobsters, crabs, sand hoppers, brine shrimps, and water fleas. The barnacle is different from all related forms, however, in its fixed and sedentary existence. When and how it assumed such a way of life is one of the riddles of zoology, the transitional forms having been lost somewhere in the mists of the past. Some faint suggestions of a similar manner of life—the waiting in a fixed place for the sea to bring food—are found among the amphipods, another group of crustaceans. Some of these spin little webs or cocoons of natural silk and

Feeding barnacles

Mass of barnacles, showing individuals

seaweed fibers; though remaining free to come and go they spend much of their time within them, taking their food from the currents. Another amphipod, a Pacific coast species, burrows into colonies of the tunicate called the sea pork, hollowing out for itself a chamber in the tough, translucent substance of its host. Lying in this excavation, it draws currents of sea water over its body and extracts the food.

However the barnacle became what it is, its larval stages clearly proclaim its crustacean ancestry, although early zoologists who looked at its hard shells labeled it a mollusk. The eggs develop inside the parent's shell and presently hatch into the sea in milky clouds of larvae. (The British zoologist Hilary Moore, after studying barnacles on the Isle of Man, estimated a yearly production of a million million larvae from a little over half a mile of shore.) Larval life lasts about three months in the rock barnacle, with several molts and transformations of form. At first the larva, a little swimming creature called a nauplius, is indistinguishable from the larva of all other crustaceans. It is nourished by large globules of fat that not only feed it but keep it near the surface. As the fat globules dwindle, the larva begins to swim at lower water levels. Eventually it changes shape, acquires a pair of shells, six pairs of swimming legs, and a pair of antennae tipped with suckers. This "cypris" larva looks much like the adults of another group of crustaceans, the ostracods. Finally, guided by instinct to yield to gravity and to avoid light, it descends to the bottom ready to become an adult.

No one knows how many of the baby barnacles riding shoreward on the waves makes a safe landing, how many fail in the quest for a clean, hard substratum. The settling down of a barnacle larva is not a haphazard process, but is performed only after a period of seeming deliberation. Biologists who have observed the act in the laboratory say the larvae "walk" about on the substratum for as long as an hour, pulling themselves along by the adhesive tips of the antennae, testing and rejecting many possible sites before they make a final choice. In nature they probably drift along in the currents for many days, coming down, examining the bottom at hand, then drifting on to another.

What are the conditions this infant creature requires? Probably it finds rock surfaces that are rough and pitted better than very smooth ones; probably it is repelled by a slimy film of microscopic plants, or even sometimes by the presence of hydroids or large algae. There is some reason to believe it may be drawn to existing colonies of barnacles perhaps through mysterious chemical attraction, detecting substances released by the adults and following these paths to the colony. Somehow, suddenly and irrevocably, the choice is made and the young barnacle cements itself to the chosen surface. Its tissues undergo a complete and drastic reorganization comparable to the metamorphosis of the larval butterfly. Then from an almost shapeless mass, the rudiments of the shell appear, the head and appendages are molded, and within twelve hours the complete cone of the shell, with all its plates delineated, has been formed.

Within its cup of lime the barnacle faces a dual growth problem. As a crustacean enclosed in a chitinous shell, the animal itself must periodically shed its unyielding skin so that its body may enlarge. Difficult as it seems, this feat is successfully accomplished, as I am reminded many times each summer. Almost every container of sea water that I bring up from the shore is flecked with white semitransparent objects, gossamer-fine, like the discarded garments of some very small fairy creature. Seen under the microscope, every detail of structure is perfectly represented. Evidently the barnacle accomplishes its withdrawal from the old skin with incredible neatness and thoroughness. In the little cellophane-like replicas I can count the joints of the appendages; even the bristles, growing at the bases of the joints, seem to have been slipped intact out of their casings.

The second problem is that of enlarging the hard cone to accommodate the growing body. Just how this is done no one seems to be sure, but probably there is some chemical secretion to dissolve the inner layers of the shell as new material is added on the outside.

Unless its life is prematurely ended by an enemy, a rock barnacle is likely to live about three years in the middle and lower tidal zones, or five years near the upper tidal levels. It can withstand high temperatures as rocks absorb the heat of the summer sun. Winter cold in itself is not harmful, but grinding ice may scrape the rocks clean. The pounding of the surf is part of the normal life of a barnacle; the sea is not its enemy.

When, through the attacks of fish, predatory worms, or snails, or through natural causes, the barnacle's life comes to an end, the shells remain attached to the rocks. These become shelter for many of the minute beings of the shore. Besides the baby periwinkles that regularly live there, the little tide-pool insects often hurry into these shelters if caught by the rising tide. And lower on the shore, or in tide pools, the empty shells are likely to house young anemones, tube worms, or even new generations of barnacles.

The chief enemy of the barnacle on these shores is a brightly colored carnivorous marine snail, the dog whelk. Although it preys also on mussels and even occasionally on periwinkles, it seems to prefer barnacles to all other food, probably because they are more easily eaten. Like all snails, the whelk possesses a radula. This is not used, periwinkle fashion, to scrape the rocks, but to drill a hole in any hard-shelled prey. It can then be pushed through the hole it has made, to reach and consume the soft parts within. To devour a barnacle, however, the whelk need only envelop the cone within its fleshy foot and force the valves open. It also produces a secretion that may have a narcotic effect. This is a substance called purpurin. In ancient times the secretion of a related snail in the Mediterranean was the source of the dye Tyrian purple. The pigment is an organic compound of bromine that changes in air to form a purple coloring matter.

Although violent surf excludes them, the dog whelks appear in numbers on most open shores, working up high into the zone of the barnacles and mussels. By their voracious feeding they may actually alter the balance of life on the shore. There is a story, for example, about an area where the whelks had reduced the number of barnacles so drastically that mussels came in to fill the vacant niche. When the whelks could find no more barnacles they moved over to the mussels. At first they were clumsy, not knowing how to eat the new food. Some spent futile days boring holes in empty shells; others climbed into empty shells and bored

Dog whelks feeding on barnacles

from inside. In time, however, they adjusted to the new prey and ate so many mussels that the colony began to dwindle. Then the barnacles settled anew on the rocks and in the end the snails returned to them.

In the middle stretches of shore and even down toward the low-tide line the whelks live under the dripping curtains of weed hanging down from the rock walls, or within the turf of Irish moss or among the flat, slippery fronds of the red seaweed, dulse. They cling to the under sides of overhanging ledges or gather in deep crevices where salt water drips from weeds and mussels, and little streams of water trickle over the floor. In all such places the whelks collect in numbers to pair and lay their eggs in straw-colored containers, each about the size and shape of a grain of wheat and tough as parchment. Each capsule stands alone, attached by its own base to the substratum, but usually they are crowded so closely together that they form a pattern or mosaic.

A snail takes about an hour to make one capsule but seldom completes more than 10 in twenty-four hours. It may produce as many as 245 in a season. Although a single capsule may contain as many as a thousand eggs, most of these are unfertilized nurse eggs that serve as food for the developing embryos. On maturing, the capsules become purple, being colored by the same chemical purpurin that is secreted by the adult. In about four months embryonic life is completed, and from 15 to 20 young dog whelks emerge from the capsule. Newly hatched young seldom if ever are found in the zone where the adults live, although the capsules are deposited and development takes place there. Apparently the waves carry the young snails down to low-tide level or below. Probably many are washed out to sea and lost, but the survivors are to be found at low water.

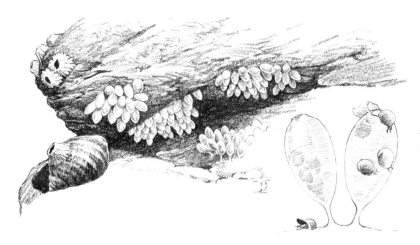

Egg capsules of the dog whelk; right: *young whelks emerging*

Rock showing various zones—black zone of algae, barnacles, rockweed, and Irish moss

They are very small—about one-sixteenth of an inch high—and are feeding on the tube worm, Spirorbis. Apparently the tubes of these worms are easier to penetrate than the cones of even very small barnacles. Not until the dog whelk is about one-fourth or three-eighths of an inch high does it migrate higher on the shore and begin feeding on barnacles.

Down in the middle sections of the shore the limpets become abundant. They appear sprinkled over the open rock surfaces, but most live numerously in shallow tide pools. A limpet wears a simple cone of shell

35

the size of a fingernail, unobstrusively mottled with soft browns and grays and blues. It is one of the most ancient and primitive types of snails, and yet the primitiveness and the simplicity are deceptive. The limpet is adapted with beautiful precision to the difficult world of the shore. One expects a snail to have a coiled shell; the limpet has instead a flattened cone. The periwinkles, which have coiled shells, are often rolled around by the surf unless they have hidden themselves securely in crevices or under weeds. The limpet merely presses its cone against the rocks and the water slides over the sloping contours without being able to get a grip; the heavier the waves, the more tightly they press it to the rocks. Most snails have an operculum to shut out enemies and keep moisture inside; the limpet has one in infancy and then discards it. The fit of the shell to the substratum is so close that an operculum is unnecessary; moisture is retained in a little groove that runs around just inside the shell, and the gills are bathed in a small sea of their own until the tide returns.

Ever since Aristotle reported that limpets leave their places on the rocks and go out to feed, people have been recording facts about their natural history. Their supposed possession of a sort of homing sense has been widely discussed. Each limpet is said to possess a "home" or spot to which it always returns. On some types of rock there may be a recognizable scar, either a discoloration or a depression, to which the contours of the shell have become precisely fitted. From this home the limpet wanders out on the high tides to feed, scraping the small algae off the rocks by licking motions of the radula. After an hour or two of feeding it returns by approximately the same path, and settles down to wait out the period of low water.

Many nineteenth-century naturalists tried unsuccessfully to discover by experiment the nature of the sense involved and the organ in which the homing sense resides, much as modern scientists try to find a physical basis for the homing abilities of birds. Most of these studies dealt with the common British limpet, Patella, and although no one could explain how the homing instinct worked, there seemed to be little doubt in anyone's mind that it did work, and with great precision.

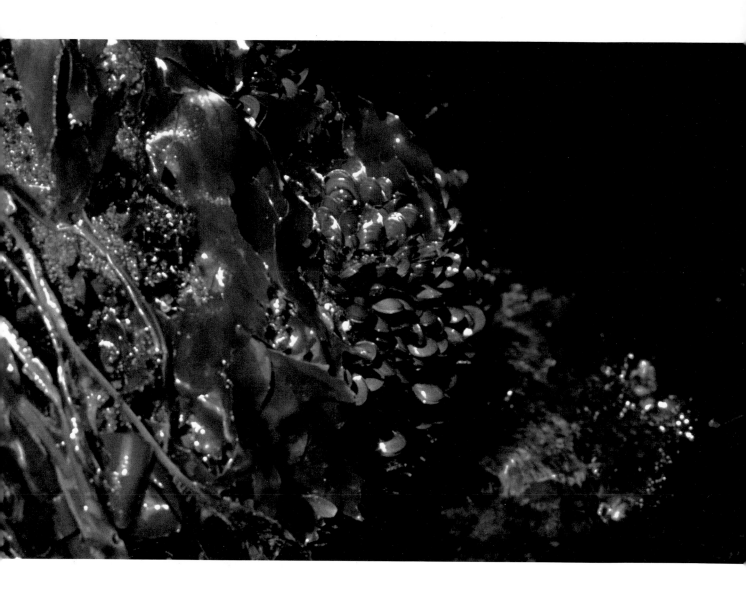

In recent years American scientists have investigated the matter with statistical methods, and some have concluded that Pacific coast limpets do not "home" very well at all. (No careful studies of homing have been made among New England limpets.) Other recent work in California, however, supports the homing theory. Dr. W. G. Hewatt marked a large number of limpets and their homes with identifying numbers. He found that on each high tide all the limpets left home, wandered about for some two and a half hours, then returned. The direction of their excursions changed from tide to tide, but they always returned to the home spot. Dr. Hewatt tried filing a deep groove across one animal's homeward path. The limpet halted on the edge of the groove and spent some time confronting this dilemma, but on the next tide it moved around the edge of the groove and returned home. Another limpet was taken about nine inches from its home and the edges of its shell were filed smooth. It was then released on the same spot. It returned to its home, but presumably the exact fit of shell to rock home had been destroyed by the filing and the next day the limpet moved about twenty-one inches away and did not return. On the fourth day it had taken up a new home and after eleven days it disappeared.

The limpet's relations with other inhabitants of the shore are simple. It lives entirely on the minute algae that coat the rocks with a slippery film, or on the cortical cells of larger algae. For either purpose, the radula is effective. The limpet scrapes the rocks so assiduously that fine particles of stone are found in its stomach; as the teeth of the radula wear away under hard use they are replaced by others, pushed up from behind. To the algal spores swarming in the water, ready to settle down and become sporelings and then adult plants, the limpets stand in the relation of enemy, since they keep the rocks scraped fairly clean where they are numerous. By this very act, however, they render a service to barnacles, making easier the attachment of their larvae. Indeed, the paths radiating out from a limpet's home are sometimes marked by a sprinkling of the starlike shells of young barnacles.

In its reproductive habits this deceptively simple little snail seems again to have defied exact observation. It seems certain, however, that the female limpet does not make protective capsules for her eggs in the fashion typical of many snails, but commits them directly to the sea. This is a primitive habit, followed by many of the simpler sea creatures. Whether the eggs are fertilized within the mother's body or while floating at sea is uncertain. The young larvae drift or swim for a time in the surface waters; the survivors then settle down on rocky surfaces and metamorphose from the larval to the adult form. Probably all young limpets are males, later transforming to females—a circumstance not at all uncommon among mollusks.

Like the animal life of this coast, the seaweeds tell a silent story of heavy surf. Back from the headlands and in bays and coves the rockweeds may grow seven feet tall; here on this open coast a seven-inch plant is a large one. In their sparse and stunted growth, the seaweed invaders of the upper rocks reveal the stringent conditions of life where waves beat heavily. In the middle and lower zones some hardy weeds have been able to establish themselves in greater abundance and profusion. These differ so greatly from the algae of quieter shores that they are almost a symbol of the wave-swept coast. Here and there the rocks sloping to the sea glisten with sheets composed of many individual plants of a curious seaweed, the purple laver. Its generic name, Porphyra, means "a purple dye." It belongs to the red algae, and although it has color variations, on the Maine coast it is most often a purplish brown. It resembles nothing so much as little pieces of brown transparent plastic cut out of someone's raincoat. In the thinness of its fronds it is like the sea lettuce, but there is a double layer of tissue, suggesting a child's rubber balloon that has collapsed so that the opposite walls are in contact. At the stem of the "balloon" Porphyra is attached strongly to the rocks by a cord of interwoven strands—hence its specific name, "umbilicalis." Occasionally it is attached to barnacles and very rarely it grows on other algae instead of directly on hard surfaces. When exposed at ebb tide under a hot sun, the laver may dry to brittle, papery layers, but the return of the sea restores the elastic nature of the plant, which, despite its seeming delicacy, allows it to yield unharmed to the push and pull of waves.

Purple laver

Down at the lower tidal levels is another curious weed—Leathesia, the sea potato. It grows in roughly globular form, its surface seamed and drawn into lobes, forming fleshy, amber-colored tubers that are any size up to an inch or two in diameter. Usually it grows around the fronds of moss or another seaweed, seldom if ever attaching directly to the rocks.

The lower rocks and the walls of low tide pools are thickly matted with algae. Here the red weeds largely supplant the browns that grow higher up. Along with Irish moss, dulse lines the walls of the pools, its thin, dull red fronds deeply indented so that they bear a crude resemblance to the shape of a hand. Minute leaflets sometimes haphazardly attached along the margins create a strangely tattered appearance. With the water withdrawn, the dulse mats down against the rocks, paper-thin layers piled one upon another. Many small starfish, urchins, and mollusks live within the dulse, as well as in the deeper growth of Irish moss.

Dulse is one of the algae that have a long history of usefulness to man, as food for himself and his domestic animals. According to an old book on seaweeds, it used to be said in Scotland that "he who eats of the Dulse of Guerdie and drinks of the wells of Kildingie will escape all maladies except black death." In Great Britain, cattle are fond of it and sheep wander down into the tidal zone at low water in search of it. In Scotland, Ireland, and Iceland people eat dulse in various ways, or dry it and chew it like tobacco; even in the United States, where such foods are usually ignored, it is possible to buy dulse fresh or dried in some coastal cities.

In the very lowest pools the Laminarias begin to appear, called variously the oarweeds, devil's aprons, sea tangles, and kelps. The Laminarias belong to the brown algae, which flourish in the dimness of deep waters and polar seas. The horsetail kelp lives below the tidal zone with others of the group, but in deep pools also comes over the threshold, just above the line of the lowest tides. Its broad, flat, leathery frond is frayed into long ribbons, its surface is smooth and satiny, and its color richly, glowingly brown.

The water in these deep basins is icy cold, filled with dusky, swaying plants. To look into such a pool is to behold a dark forest, its foliage like

Sea squirts

the leaves of palm trees, the heavy stalks of the kelps also curiously like the trunks of palms. If one slides his fingers down along such a stalk and grips just above the holdfast, he can pull up the plant and find a whole microcosm held within its grasp.

One of these laminarian holdfasts is something like the roots of a forest tree, branching out, dividing, subdividing, in its very complexity a measure of the great seas that roar over this plant. Here, finding secure attachment, are plankton-strainers like mussels and sea squirts. Small starfish and urchins crowd in under the arching columns of plant tissue. Predacious worms that have foraged hungrily during the night return with the daylight and coil themselves into tangled knots in deep recesses and dark, wet caverns. Mats of sponge spread over the holdfasts, silently, endlessly at their work of straining the waters of the pool. One day a larval bryozoan settles here, builds its tiny shell, then builds another and another, until a film of frosty lace flows around one of the rootlets of the seaweed. And above all this busy community, and probably unaffected by it, the brown ribbons of the kelp roll out into the water, the plant living its own life, growing, replacing torn tissues as best it may, and in season sending clouds of reproductive cells streaming into the water. As for the fauna of the holdfasts, the survival of the kelp is their survival. While it stands firm their little world holds intact; if it is torn away in a surge of stormy seas, all will be scattered and many will perish with it.

Kelp's holdfast

Among the animals almost always inhabiting the holdfasts of the tide-pool kelps are the brittle stars. These fragile echinoderms are well named, for even gentle handling is likely to cause them to snap off one or more arms. This reaction may be useful to an animal living in a turbulent world, for if an arm is pinned down under a shifting rock, the owner can break it off and grow a new one. Brittle stars move about rapidly, using their flexible arms not only in locomotion, but also to capture small worms and other minute sea life and carry them to their mouths.

The scale worms also belong to the holdfast community. Their bodies are protected by a double row of plates forming armament over the back. Under these large plates is an ordinary segmented worm, bearing laterally projecting tufts of golden bristles on each segment. There is a suggestion of primitiveness in the armor plate that is reminiscent of the quite unrelated chitons. Some of the scale worms have developed interesting relations with their neighbors. One of the British species always lives with burrowing animals, although it may change associates from time to time. When young, it lives with a burrowing brittle star, probably stealing its food. When older and larger, it moves into the burrow of a sea cucumber or the tube of the much larger, plumed worm, Amphitrite.

Often the holdfast grips one of the large horse mussels, which have heavy shells and may be four or five inches long. The horse mussel lives only in the deep pools or farther offshore; it is never found in the upper zones with the smaller blue mussel, and it occurs only on or among rocks, where its attachment is relatively secure. Sometimes it constructs a small nest or den as a refuge, using tough byssal threads spun in typical mussel fashion, with pebbles and shell fragments matted among the strands.

A small clam common in laminarian holdfasts is the rockborer, which some English writers call the "red-nose" because of its red siphons. Ordinarily it is a boring form, living in cavities it excavates in limestone, clay, or concrete. Most of the New England rocks are too hard for boring, and so on this coast the clam lives in crusts of coralline algae or among the holdfasts of the kelp. On British coasts it is said to bore rocks that resist mechanical drills. And it does so without recourse to the chemical secretions some borers use, working entirely by repeated and endless mechanical abrasion with its sturdy shell.

44

Brittle star

The smooth and slippery fronds of the kelps support other populations, less abundant and less varied than those of the holdfasts. On the flat blades of the oarweeds, as well as on rock faces and under ledges, the golden-star tunicate, Botryllus, lays its spangled mats. Over a field of dark green gelatinous substance are sprinkled the little golden stars that mark the position of clusters of individual tunicates. Each starry cluster may consist of three to a dozen individual animals radiating around a central point; many clusters go to make up this continuous, encrusting mat, which may be six to eight inches long.

Beneath the surface beauty there is marvelous complexity of structure and function. Over each star there are infinitesimal disturbances in the water—little currents funneling down, one to each point of the star, and there being drawn in through a small opening. One heavier, outward-moving current emerges from the center of the cluster. The indrawn currents bring in food organisms and oxygen, and the outflowing current carries away the metabolic waste products of the group.

At first glance a colony of Botryllus may seem no more complex than a mat of encrusting sponge. In actual fact, however, each of the individuals comprising the colony is a highly organized creature, in structure almost identical with such solitary ascidians as the sea grape and the sea vase, found so abundantly on wharves and sea walls. The individual Botryllus, however, is only one-sixteenth to one-eighth of an inch long.

One of these entire colonies, comprising perhaps hundreds of star clusters (and so perhaps a thousand or more individuals), may arise from a single fertilized ovum. In the parent colony, eggs are formed early in the summer, are fertilized, and begin their development while remaining within the parental tissues. (Each individual Botryllus produces both eggs and sperms, but since in any one animal they mature at different times, cross fertilization is insured, the spermatozoa being carried in the sea water and drawn in along with the water currents.) Presently the parent releases minute larvae shaped like tadpoles, with long, swimming tails. For perhaps an hour or two such a larva drifts and swims, then settles down on some ledge or weed and makes itself fast. Soon the tissues of the tail are

Golden-star tunicate

absorbed and all suggestion of ability to swim is lost. Within two days the heart begins to beat in that curious tunicate rhythm—first driving the blood in one direction, pausing briefly, then reversing the direction of the flow. After nearly a fortnight, this small individual has completed the formation of its own body and begins to bud off other individuals. These, in turn, bud off others. Each new creature has its separate opening for the intake of water, but all retain connections with a central vent for the outflow of wastes. When the individuals clustered around this common opening become too crowded, one or more newly formed buds are pushed out into the surrounding mat of gelatinous tissue, where they begin new star clusters. In this way the colony spreads.

The intertidal zone is sometimes invaded by a deep-water laminarian, the sea colander. It is a representative of those brown seaweeds that flourish in the cold waters of the Arctic, and has come down from Greenland as far as Cape Cod. Its appearance is strikingly different from that of the sea moss and horsetail kelp among which it sometimes appears. The wide frond is pierced by innumerable perforations; these are foreshadowed in the young plant by conical papillae, which later break through to form the perforations.

Beyond the rims of the lowest pools, growing on the rock walls that slope away steeply into deep water, is another laminarian seaweed, Alaria,

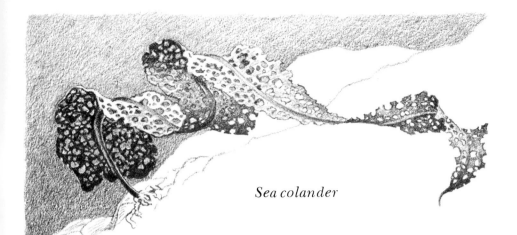

Sea colander

the winged kelp, called the murlin in Great Britain. Its long, ruffled, streaming fronds rise with each surge and fall as the water pours away seaward. The fertile pinnae, in which the reproductive cells mature, are borne at the base of the frond, for in a plant so exposed to violent surf this location is safer than the tips of the main blade. (In the rockweeds, living higher on the shore and less subject to savage wave action, the reproductive cells are formed at the tips of the fronds.) Almost more than any of the other seaweeds, Alaria is a plant conditioned to constant pounding by the waves. Standing on the outermost point that gives safe footing, one can see its dark ribbons streaming out into the water, being tugged and tossed and pounded. The larger and older plants become much frayed and worn, the margins of the blade splitting or the tip of the midrib being worn off. By such concessions the plant saves some of the strain on its holdfasts. The stipe can withstand a relatively enormous pull, but severe storms tear away many plants.

Still farther down, one can sometimes and in some places get a glimpse of the dark, mysterious forests of the kelps, where they go down into deep water. Sometimes these giant kelps are cast ashore after a storm. They have a stiff, strong stipe from which the long ribbon of the frond extends. The sea tangle or sugar kelp, *Laminaria saccharina*, has a stipe up to 4 feet long, supporting a relatively narrow frond (6 to 18 inches wide) that may extend out and upward into the sea as much as 30 feet. The

Winged kelp, Alaria

margin is greatly frilled and a powdery white substance (mannitol, a sugar) forms on the dried fronds. The long-stalked laminaria (*Laminaria longicruris*) has a stem comparable to the trunk of a small tree, being 6 to 12 feet long. The frond is up to 3 feet wide and 20 feet long, but may sometimes be shorter than the stipe.

The stands of sea tangles and long-stalked laminarias are, in their way, an Atlantic counterpart of the great submarine jungles of the Pacific, where the kelps rise like giant forest trees, 150 feet from the floor of the sea to the surface.

On all rocky coasts, this laminarian zone just below low water has been one of the least-known regions of the sea. We know little about what lives here throughout the year. We do not know whether some of the forms that disappear from the intertidal area in winter may merely move down into this zone. And perhaps some of the species we think have died out in a particular region, perhaps because of temperature changes, have gone down among the laminarias. The area is obviously difficult to explore, with heavy seas breaking there most of the time. Such an area on the west coast of Scotland was, however, explored by helmet divers working with the British biologist J. A. Kitching. Below the zone occupied by Alaria and the horsetail kelp, from about two fathoms below low water and beyond, the divers moved through a dense forest of the larger laminarias. From the vertical stipes an immense canopy of fronds was spread above their heads. Although the sun shone brightly at the surface, the divers were almost in darkness as they pushed through this forest. Between three and six fathoms below low water of the spring tides the forest opened out; so that the men could walk between the plants without great difficulty. There the light was stronger, and through misty water they could see this more open "park" extending farther down the sloping floor of the sea. Among the holdfasts and stipes of the laminarias, as among the roots and trunks of a terrestrial forest, was a dense undergrowth, here formed of various red algae. And as small rodents and other creatures have their dens and runways under the forest trees, so a varied and abundant fauna lived on and among the holdfasts of the great seaweeds.

49

In quieter waters, protected from the heavy surf of coasts that face the open ocean, the seaweeds dominate the shore, occupying every inch of space that the conditions of tidal rise and fall allow them and by the sheer force of abundant and luxuriant growth forcing other shore inhabitants to accommodate to their pattern.

Although the same bands of life are spread between the tide lines whether the coast be open or sheltered, in their relative development the zones vary greatly on the two types of shore.

Above the high-tide line there is little change and on the shores of bays and estuaries, as elsewhere, the microplants blacken the rocks and the lichens come down and tentatively approach the sea. Below high water of spring tides, pioneering barnacles trace occasional white streaks in token occupation of the zone they dominate on open coasts. A few periwinkles graze on the upper rocks. But on sheltered coasts the whole band of shore marked out by the tides of the moon's quarters is occupied by a swaying submarine forest, sensitive to the movements of the waves and the tidal currents. The trees of the forest are the large seaweeds known as the rockweeds or sea wracks, stout of form and rubbery of texture. Here all other life exists within their shelter—a shelter so hospitable to small things needing protection from drying air, from rain, and from the surge of the running tides and the waves, that the life of these shores is incredibly abundant.

When covered at high tide, the rockweeds stand erect, rising and swaying with a life borrowed from the sea. Then, to one standing at the edge of a flooding tide, the only sign of their presence may be a scattering of dark patches on the water close inshore, where the tips of the weeds reach up to the surface. Down below those floating tips small fishes swim, passing between the weeds as birds fly through a forest, sea snails creep along the fronds, and crabs climb from branch to branch of the swaying plants. It is a fantastic jungle, mad in a Lewis Carroll sort of way. For what proper jungle, twice every twenty-four hours, begins to sag lower and lower and finally lies prostrate for several hours, only to rise again? Yet this is precisely what the rockweed jungles do. When the tide has

retreated from the sloping rocks, when it has left the miniature seas of the tide pools, the rockweeds lie flat on the horizontal surfaces in layer above layer of sodden, rubbery fronds. From the sheer rock faces they hang down in a heavy curtain, holding the wetness of the sea, and nothing under their protective cover ever dries out.

By day the sunlight filters through the jungle of rockweeds to reach its floor only in shifting patches of shadow-flecked gold; by night the moonlight spreads a silver ceiling above the forest—a ceiling streaked and broken by the flowing tide streams; beneath it the dark fronds of the weeds sway in a world unquiet with moving shadows.

But the flow of time through this submarine forest is marked less by the alternation of light and darkness than by the rhythm of the tides. The lives of its creatures are ruled by the presence or absence of water; it is not the fall of dusk or the coming of dawn but the turn of the tide that brings transforming change to their world.

As the tide falls the tips of the weeds, lacking support, float out horizontally across the surface. Then the cloud shadows darken and a deepening gloom settles over the floor of the forest. As the overlying layer of water thins and gradually drains away, the weeds, still stirring, still responsive to each pulsation of the tide, drift closer to the rock floor and finally lie prostrate upon it, all their life and movement in abeyance.

By day an interval of quiet settles over the jungles of the land, when the hunters lie in their dens, and the weak and the slow hide from the daylight; so on the shore a waiting lull comes with every ebbing of the tide.

The barnacles furl their nets and swing shut the twin doors that exclude the drying air and hold within the moisture of the sea. The mussels and the clams withdraw their feeding tubes or siphons and close their shells. Here and there a starfish, having invaded the forest from below on the previous high tide and incautiously lingered, still clasps a mussel within its sinuous arms, gripping the shells with the sucker-tipped ends of scores of slender tube feet. Pushing under and among the horizontal fronds of the weed, as a man would make his way with difficulty

through trees blown down by a storm, a few crabs are active, digging their little slanting pits to expose the clams buried in the mud. Then they crack away pieces of shell with their heavy claws, while they hold the clam in the tips of the walking legs.

A few hunters and scavengers come down from the upper tidelands. The little gray-cloaked tide-pool insect, Anurida, wanders down from the upper shore and scurries over the rock floor, hunting out mussels with gaping shells, or dead fish, or fragments of crabs left by gulls. Crows walk about over the weeds; they sort them over strand by strand until they find a periwinkle hidden in the weed, or clinging to a rock that lies under the sodden cloak of the algae. Then the crow holds the shell in the strong toes of one foot, while with its beak it deftly extracts the snail.

The pulse of the returning tide at first beats gently. The advance during the beginning of the six-hour rise to high-water mark is slow, so that in two hours only a quarter of the intertidal zone has been covered. Then the pace of the water quickens. For the next two hours the tidal currents are stronger and the rising waters advance twice as far as in the first period; then again the tide slackens its pace for a leisurely advance over the upper shore. The rockweeds, covering the middle band of shore, receive the shock of heavier waves than the relatively bare shore above, yet their cushioning effect is so great that the animals that cling to them or live on the rock floor below them are far less affected by the surf than those of the upper rocks, or those of the zone below which experience all the heavy drag from the backwash of waves that break as the tide is advancing rapidly over the middle shore.

Darkness brings the jungles of the land to life, but the night of the rockweed jungles is the time of the rising tide, when water pours in under the masses of weed, stirring out of their low-tide quiescence all the inhabitants of this forest.

As the water from the open sea floods the floor of the weed jungles, shadows flicker again above the ivory cones of the barnacles as the almost invisible nets reach out to gather what the tide has brought. The shells of clams and mussels again open slightly and little vortices of water are

56

drawn down, funneling into the complex straining mechanisms within the shellfish all the little spheres of marine vegetables that are their food.

Nereid worms emerge from the mud and swim off to other hunting grounds; if they are to reach them they must elude the fishes that have come in with the tide, for on the flood tide the rockweed forests become one with the sea and with its hungry predators.

Shrimp flicker in and out through the open spaces of the forest; they seek small crustaceans, baby fish, or minute bristle worms, but in their turn are pursued by following fish. Starfish move up from the great meadows of sea moss lower on the shore, hunting the mussels that grow on the floor of the forest.

The crows and the gulls are driven out of the tidelands. The little gray, velvet-cloaked insects move up the shore or, finding a secure crevice, wrap themselves in a glistening blanket of air to wait for the falling of the tide.

The rockweeds that create this intertidal forest are descendants of some of the earth's most ancient plants. Along with the great kelps lower on the shore, they belong to the group of brown seaweeds, in which the chlorophyll is masked by other pigments. The Greek name for the brown algae—the *Phaeophyceae*—means "the dusky or shadowy plants." According to some theories, they arose in that early period when the earth was still enveloped in heavy clouds and illuminated only by feeble rays of sunlight. Even today the brown seaweeds are plants of dim and shadowed places—the deep submarine slopes where giant kelps form dusky jungles and the dark rock ledges from which the oarweeds send their long ribbons streaming into the tides. And the rockweeds that grow between the tide lines do so on northern coasts, visited often by cloud and fog. Their rare invasions of the sunny tropics are accomplished under a protective cover of deep water.

The brown algae may have been the first of the sea plants to colonize the shore. They learned to adjust themselves to alternating periods of submersion and exposure on ancient coastlines swept by strong tides; they came as close to a land existence as they could without actually leaving the tidal zone.

One of the modern rockweeds, the channeled wrack of European shores, lives at the extreme upper edge of the tidelands. In some places its only contact with the sea is an occasional drenching with spray. In sun and air its fronds become blackened and crisp so that one would think it had surely been killed, but with the return of the sea its normal color and texture are restored.

The channeled wrack does not grow on the American Atlantic coast, but there a related plant, the spiral wrack, comes almost as far out of the sea. It is a weed of low growth, whose short sturdy fronds end in turgid, rough-textured swellings. Its heaviest growth is above the high-water mark of the neap tides, so of all the rockweeds it lives closest inshore or nearest the water line of exposed ledges. Although it spends nearly three-fourths of its life out of water, it is a true seaweed and its splashes of orange-brown color on the upper shore are a symbol of the threshold of the sea.

These plants, however, are but the outlying fringe of the intertidal forest, which is an almost pure stand of two other rockweeds—the knotted wrack and the bladder wrack. Both are sensitive indicators of the force of the surf. The knotted wrack can live in profusion only on shores protected from heavy waves, and in such places is the dominant weed. Back from the headlands, on the shores of bays and tidal rivers where surf and tidal surge are subdued by remoteness from the open sea, the knotted wrack may grow taller than the tallest man, though its fronds are slender as straws. The long surge of the swell in sheltered water places no great strain on its elastic strands. Swellings or vesicles on the main stems or fronds contain oxygen and other gasses secreted by the plant; these act as buoys when the weeds are covered by the tide. The bladder wrack has greater tensile strength and so can endure the sharp tugging and pulling of moderately heavy surf. Although it is much shorter than the knotted wrack it also needs the help of air bladders to rise in the water. In this species the bladders are paired, one of each pair on either side of the strong midrib; the bladders, however, may fail to develop where the plants are subjected to much pounding by surf, or when they grow at the lower levels of the tidal zone. At some seasons the ends of the branches of this

59

wrack swell into bulbous, almost heart-shaped structures; from these the reproductive cells are liberated.

The sea wracks have no roots, but instead grip the rocks by means of a flattened, disc-like expansion of their tissues. It is almost as though the base of each weed melted a little, spreading over the rock and then congealing, thereby creating a union so firm that only the thundering seas of a very heavy storm, or the grinding of shore ice, can tear away the plants. The seaweeds do not have a land plant's need of roots to extract minerals from the soil, for they are bathed almost continously by the sea and so live within a solution of all the minerals they need for life. Nor do they need the rigid supporting stem or trunk by which a land plant reaches upward into sunlight—they have only to yield themselves to the water. And so their structure is simple—merely a branching frond arising from the holdfast, with no division into roots and stems and leaves.

Looking at the prostrate, low-tide forests of the rockweeds that cover the shore with a many-layered blanket, one would suppose that the plants must spring from every available inch of rock surface. But actually the forest, when it rises and comes to life with the flooding tide, is fairly open and sprinkled with clearings. On my own shore in Maine, where the tides rise and fall over a wide expanse of intertidal rock, and the knotted wrack spreads its dark blanket between the high and low waters of the neap tides, the areas of open rock around the holdfast of each plant are sometimes as much as a foot in diameter. From the middle of such a clearing the plant rises, its fronds dividing repeatedly, until the upper branchings extend out over an area several feet across.

Far below, at the base of the fronds that swing with the undulation of the passing waves, the rocks are stained with vivid hues, painted in crimson and emerald by the activities of sea plants so minute that even in their thousands they seem but part of the rock, a surface revelation of jewel tones within. The green patches are growths of one of the green algae. The individual plants are so small that only a strong lens could reveal their identity—lost, as individual blades of grass are lost in the lush

61

expanse of a meadow, in the spreading verdant stain created by the mass. Amid the green are other patches of a rich and intensely glowing red, and again the growth is not separable from the mineral floor. It is a creation of one of the red seaweeds, a form that secretes lime in thin and closely adhering crusts over the rocks.

Against this background of glowing color the barnacles stand out with sharp distinctness, and in the clear water that pours through the forest like liquid glass, their cirri flicker in and out—extending, grasping, withdrawing, taking from the inpouring tides those minute atoms of life that our eyes cannot see. Around the bases of small wave-rounded boulders the mussels lie as though at anchor, held by gleaming lines spun by their own tissues. Their paired blue shells stand a little apart, the space between them revealing pale brown tissues with fluted edges.

Some parts of the forest are less open. In these the clumps of rockweeds rise from a short turf or undergrowth consisting chiefly of the flat fronds of Irish moss, with sometimes dark mats of another plant with the texture of Turkish toweling. And like a tropical jungle with its orchids, this sea forest has the counterpart of airplants in the epiphytic tufts of a red seaweed that grows on the fronds of the knotted wrack. Polysiphonia seems to have lost—or perhaps it never had—the ability to attach directly to the rocks and so its dark red balls of finely divided fronds cling to the wracks, and by them are lifted up into the water.

In the areas between the rocks and under loose boulders a substance that is neither sand nor mud has accumulated. It consists of minute and water-ground bits of the remains of sea creatures—the shells of mollusks, the spines of sea urchins, the opercula of snails. Clams live in pockets of this soft substance, digging down until they are buried to the tips of their siphons. Around the clams the mud is alive with ribbon worms, thin as

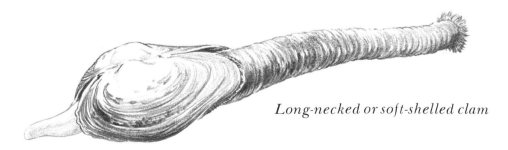

Long-necked or soft-shelled clam

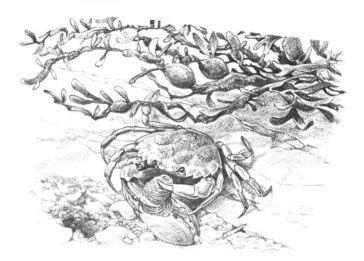

Green crab eating clam

threads, scarlet of color, each a small hunter searching out minute bristle worms and other prey. Here also are the nereids, given the Latin name for sea nymph because of their grace and iridescent beauty. The nereids are active predators that leave their burrows at night to search for small worms, crustaceans, and other prey. In the dark of the moon certain species gather at the surface in immense spawning swarms. Curious legends have become associated with them. In New England the so-called clam worm, *Nereis virens*, often takes shelter in empty clam shells. Fishermen, accustomed to finding it thus, believe it is the male clam.

Crabs of thumbnail size live in the weed and come down to hunt in these areas. They are the young of the green crab; the adults live below the tide lines on this shore except when they come into the shelter of the weeds to molt. The young crabs search the mud pockets, digging out pits and probing for clams that are about their own size.

Clams, crabs, and worms are part of a community of animals whose lives are closely interrelated. The crabs and the worms are the active predators, the beasts of prey. The clams, the mussels, and the barnacles are the plankton feeders, able to live sedentary lives because their food is brought to them by each tide. By an immutable law of nature, the plankton feeders as a group are more numerous than those that prey on them. Besides the clams and other large species, the rockweeds shelter thousands of small beings, all of them busy with filtering devices of varying design, straining out the plankton of each tide. There is, for example, a small, plumed worm called Spirorbis. Seeing it for the first time, one would certainly say that it is no worm, but a snail, for it is a tube-builder, having learned some feat of chemistry that allows it to

Coiled tubes of the worm Spirorbis

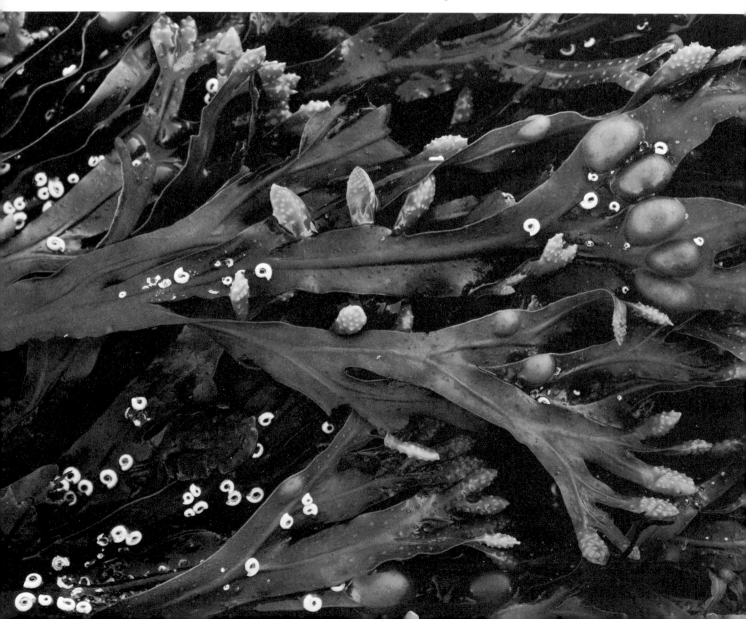

secrete about itself a calcareous shell or tube. The tube is not much larger than the head of a pin and is wound in a flat, closely coiled spiral of chalky whiteness, its form strongly suggesting some of the land snails. The worm lives permanently within the tube, which is cemented to weed or rock, thrusting out its head from time to time to filter food animals through the fine filaments of its crown of tentacles. These exquisitely delicate and filmy tentacles serve not only as snares to entangle food but as gills for breathing. Among them is a structure like a long-stemmed goblet; when the worm draws back into its tube the goblet or operculum closes the opening like a neatly fitted trap door.

The fact that the tube worms have managed to live in the intertidal zone for perhaps millions of years is evidence of a sensitive adjustment of their way of life, on the one hand to conditions within the surrounding world of the rockweeds, on the other to vast tidal rhythms linked with the movements of earth, moon, and sun.

In the inmost coils of the tube are little chains of beads wrapped in cellophane—or so they appear. There are about twenty beads in a chain. The beads are developing eggs. When the embryos have developed into larvae, the cellophane membranes rupture and the young are sent forth into the sea. By keeping the embryonic stages within the parental tube Spirorbis protects its young from enemies and assures that the infant worms will be in the intertidal zone when they are ready to settle. Their period of active swimming is short—at most an hour or so, and well contained within a single rising or falling of the tide. They are stout little creatures with bright red eye spots; perhaps the larval eyes help in locating a place for attachment but in any event they degenerate soon after the larva settles.

In the laboratory, under my microscope, I have watched the larvae swimming about busily, all their little bristles whirring, then sometimes descending to the glass floor of their dish to bump it with their heads. Why and how do the infant tube worms settle in the same sort of place their ancestors chose? Apparently they make many trials, reacting more favorably to smooth surfaces than to rough, and displaying a strong instinct of gregariousness that leads them to settle by preference where others of their kind are already established. These tendencies help to keep the tube worms within their comparatively restricted world. There is also a response, not to familiar surroundings, but to cosmic forces. Every fortnight, on the moon's quarter, a batch of eggs is fertilized and taken into the brood chamber to begin its development. And at the same time the larvae that have been made ready during the previous fortnight are expelled into the sea. By this timing—this precise synchronizing with the phases of the moon—the release of the young always occurs on a neap tide, when neither the rise nor the fall of the water is of great extent, and even for so small a creature the chances of remaining within the rockweed zone are good.

Sea snails of the periwinkle tribe inhabit the upper branches of the weeds at high tide or take shelter under them when the tide is out. The orange and yellow and olive-green colors of their smoothly rounded, flat-topped shells suggest the fruiting bodies of the rockweeds, and perhaps the resemblance is protective. The smooth periwinkle, unlike the rough, is still an animal of the sea; the salty dampness it requires is provided by the wet and dripping fronds of the seaweeds when the tide is out. It lives by scraping off the cortical cells of the algae, seldom if ever descending to the rocks to feed on the surface film as related species do. Even in its spawning habits the smooth periwinkle is a creature of the rockweeds. There is no shedding of eggs into the sea, no period of juvenile drifting in the currents. All the stages of its life are lived in the rockweeds—it knows no other home.

Curious about the early stages of this abundant snail, I have gone down into my own rockweed forests on the summer low tides to search for them. Sorting over the prostrate wrack, examining its long strands for some signs of what I sought, I have occasionally been rewarded by discovering transparent masses of a substance like tough jelly, tightly adhering to the fronds. They averaged perhaps a quarter-inch long and half as wide. Within each mass I could see the eggs, round as bubbles, dozens of them embedded in the confining matrix. One such egg mass that

Periwinkle in rockweed

Kelp

I carried to the microscope contained a developing embryo within the membranes of each egg. They were clearly molluscan, but so undifferentiated that I could not have said what mollusk lay nascent within. In the cold waters of its home, about a month would intervene from the egg to the hatching stage, but in the warmer temperatures of the laboratory the remaining days of development were reduced to hours. The following day each sphere contained a tiny baby periwinkle, its shell completely formed, apparently ready to emerge and take up its life on the rocks. How do they hold their places there, I wondered, as the weeds sway in the tides and occasional storms send waves pounding in over the shore? Later in the summer there was at least a partial answer. I noticed that many of the air vesicles of the wracks bore little perforations, as though they had been chewed or punctured by some animal. I slit some of these vesicles carefully so that I might look inside. There, secure in a green-walled chamber, were the babies of the smooth periwinkle—from two to half a dozen of them sharing the refuge of a single vesicle, secure alike against storms and enemies.

Down near the low water of the neap tides the hydroid Clava spreads its velvet patches on the fronds of the knotted wrack and the bladder wrack. Rising from its point of attachment like a plant from its root clump, each cluster of tubular animals looks like nothing so much as a spray of delicate flowers, shading from pink to rose and fringed with petal-like tentacles, all nodding in the water currents as woodland flowers nod in a gentle wind. But the swaying movements are purposive ones by which the hydroid reaches into the currents for food. In its way it is a voracious little jungle beast, all its tentacles studded with batteries of stinging cells that can be shot into its victims like poisoned arrows. When, in their ceaseless movements, the tentacles come into contact with a small crustacean or worm or the larva of some sea creature, a shower of darts is released; the prey animal becomes paralyzed and is seized and conveyed to the mouth by the tentacles.

Each of these colonies now established on the wracks came from a little swimming larva that once settled there, shed the hairy cilia by which

Hydroid Clava

it swam, attached itself, and began to elongate into a little plantlike being. A crown of tentacles formed at its free end. In time, from the base of the tubular creature, a seeming root, or stolon, began to creep over the rockweed, budding off new tubes, each complete with mouth and tentacles. So all the numerous individuals of the colony originated in a single fertilized ovum that yielded the wandering larva.

In season, the plantlike hydroid must reproduce, but by a strange circumstance it cannot itself yield the germ cells that would give rise to new larvae, for it can reproduce only non-sexually, by budding. So there is a curious alternation of generations, found again and again in many members of the large coelenterate group to which the hydroids belong, by which no individual produces offspring that resemble itself, but each is like the grandparental generation. Just below the tentacles of an individual Clava the buds of the new generation are produced—the alternate generation that intervenes between colonies of hydroids. They are pendent clusters shaped like berries. In some species the berries, or medusa buds, would drop from the parent and swim away—tiny, bell-shaped things like minute jellyfish. Clava, however, does not release its medusae but keeps them attached. Pink buds are male medusae; purple ones are female. When they are mature, each sheds its eggs or sperm into the sea. When fertilized,

72

the eggs begin to divide and through their development yield the little protoplasmic threads of larvae, which swim off through unknown waters to found some distant colonies.

During many days of midsummer, the incoming tides bring the round opalescent forms of the moon jellies. Most of these are in the weakened condition that accompanies the fulfillment of their life cycle; their tissues are easily torn by the slightest turbulence of water, and when the tide carries them in over the rockweeds and then withdraws, leaving them there like crumpled cellophane, they seldom survive the tidal interval.

Each year they come, sometimes only a few at a time, sometimes in immense numbers. Drifting shoreward, their silent approach is unheralded even by the cries of sea birds, who have no interest in the jellyfish as food, for their tissues are largely water.

During much of the summer they have been drifting offshore, white gleams in the water, sometimes assembling in hundreds along the line of meeting of two currents, where they trace winding lines in the sea along these otherwise invisible boundaries. But toward autumn, nearing the end of life, the moon jellies offer no resistance to the tidal currents, and almost every flood tide brings them in to the shore. At this season the adults are carrying the developing larvae, holding them in the flaps of tissue that hang from the under surface of the disc. The young are little pear-shaped creatures; when finally they are shaken loose from the parent (or freed by the stranding of the parent on the shore), they swim about in the shallow water, sometimes swarms of them together. Finally they seek bottom and each becomes attached by the end that was foremost when it swam. As a tiny plantlike growth, about an eighth of an inch high and bearing long tentacles, this strange child of the delicate moon jelly survives the winter storms. Then constrictions begin to encircle its body, so that it comes to resemble a pile of saucers. In the spring these "saucers" free themselves one after another and swim away, each a tiny jellyfish, fulfilling the alternation of the generations. North of Cape Cod these young grow to their full diameter of six to ten inches by July; they mature and produce

Moon jelly

eggs and sperm cells by late July or August; and in August and September they begin to yield the larvae that will become the attached generation. By the end of October all the season's jellyfish have been destroyed by storms, but their offspring survive, attached to the rocks near the low-tide line or on nearby bottoms offshore.

If the moon jellies are symbols of the coastal waters, seldom straying more than a few miles offshore, it is otherwise with the great red jellyfish, Cyanea, which in its periodic invasions of bays and harbors links the shallow green waters with the bright distances of the open sea. On fishing banks a hundred or more miles offshore one may see its immense bulk drifting at the surface as it swims lazily, its tentacles sometimes trailing for fifty feet or more. These tentacles spell danger for almost all sea creatures in their path and even for human beings, so powerful is the sting. Yet young cod, haddock, and sometimes other fishes adopt the great jellyfish as a "nurse," traveling through the shelterless sea under the protection of this large creature and somehow unharmed by the nettle-like stings of the tentacles.

Like Aurelia, the red jellyfish is an animal only of the summer seas, for whom the autumnal storms bring the end of life. Its offspring are the winter plantlike generation, duplicating in almost every detail the life history of the moon jelly. On bottoms no more than two hundred feet deep (and usually much less), little half-inch wisps of living tissue represent the heritage of the immense red jellyfiish. They can survive the cold and the storms that the larger summer generation cannot endure; when the warmth of spring begins to disspate the icy cold of the winter sea they will bud off the tiny discs that, by some inexplicable magic of development, grow in a single season into the adult jellyfish.

As the tide falls below the rockweeds, the surf of the sea's edge washes over the cities of the mussels. Here, within these lower reaches of the intertidal zone, the blue-black shells form a living blanket over the rocks. The cover is so dense, so uniform in its texture and composition, that often one scarcely realizes that this is not rock, but living animals. In one place the shells, unimaginable in number, are no more than a quarter of an

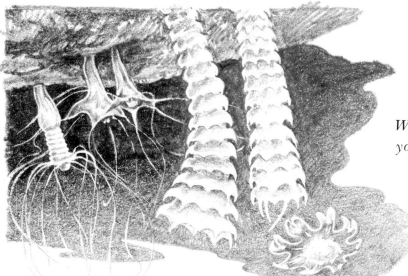

Winter stage of Aurelia, budding off young jellyfish

inch long; in another the mussels may be several times as large. But always they are packed so closely together, neighbor against neighbor, that it is hard to see how any one of them can open its shells enough to receive the currents of water that bring its food. Every inch, every hundredth of an inch of space, has been taken over by a living creature whose survival depends on gaining a foothold on this rocky shore.

The presence of each individual mussel in this crowded assemblage is evidence of the achievement of its unconscious, juvenile purpose, an expression of the will-to-live embodied in a minute transparent larva once set adrift in the sea to find its own solid bit of earth for attachment, or to die.

The setting adrift takes place on an astronomical scale. Along the American Atlantic coast the spawning season of the mussels is protracted, extending from April into September. What induces a wave of spawning at any particular time is unknown, but it seems clear that the spawning of a few mussels releases chemical substances into the water, and that these react on all mature individuals in the area and set them to pouring their eggs and milt into the sea. The female mussels discharge the eggs in a continuing, almost endless stream of short little rodlike masses—hundreds, thousands, millions of cells, each potentially an adult mussel. One large female may release up to twenty-five million at a single spawning. In quiet water the eggs drift gently to the bottom, but in the normal conditions of surf or swiftly moving currents they are at once possessed by the sea and carried away.

Simultaneously with the outflow of eggs, the water has become

cloudy with the milt poured into the water by the male mussels, the number of individual sperm cells defying all attempts at calculation. Dozens of them cluster about a single egg, pressing against it, seeking entrance. But one male cell, and one only, is successful. With the entrance of this first sperm cell, an instantaneous physical change takes place in the outer membranes of the egg, and from this moment it cannot again be penetrated by a spermatozoan.

After the union of the male and female nuclei, the division of the fertilized cell proceeds rapidly. In less than the interval between a high and a low tide, the egg has been transformed into a little ball of cells, propelling itself through the water with glittering hairs, or cilia. In about twenty-four hours, it has assumed an odd, top-shaped form that is common to the larvae of all young mollusks and annelid worms. A few days more and it has become flattened and elongated and swims rapidly by vibrations of a membrane called the velum; it crawls over solid surfaces, and senses contact with foreign objects. Its journey through the sea is far from being a solitary one; in a square meter of surface over a bed of adult mussels there may be as many as 170,000 swimming larvae.

The thin larval shell takes form, but soon it is replaced by another, double-valved as in adult mussels. By this time the velum has disintegrated, and the mantle, foot, and other organs of the adult have begun their development.

From early summer these tiny shelled creatures live in prodigious numbers in the seaweeds of the shore. In almost every bit of weed I pick up for microscopic examination I find them creeping about, exploring their world with the long tubular organ called the foot, which bears an odd resemblance to the trunk of an elephant. The infant mussel uses it to test out objects in its path, to creep over level or steeply sloping rocks or through seaweeds, or even to walk on the under side of the surface film of quiet water. Soon, however, the foot assumes a new function: it aids in the work of spinning the tough silken threads that anchor a mussel to whatever offers a solid support and insurance against being washed away in the surf.

Tide pool floor, with mussels and whelks

The very existence of the mussel fields of the low-tide zone is evidence that this chain of circumstances has proceeded unbroken to its consummation untold millions upon millions of times. Yet, for every mussel surviving upon the rocks, there must have been millions of larvae whose setting forth into the sea had a disastrous end. The system is in delicate balance; barring catastrophe, the forces that destroy neither outweight nor are outweighed by those that create, and over the years of a man's life, as over the ages of recent geologic time, the total number of mussels on the shore probably has remained about the same.

In much of this low-water area the mussels live in intimate association with one of the red seaweeds, Gigartina, a plant of low-growing, bushy form and almost cartilaginous texture. Plants and mussels unite inseparably to form a tough mat. Very small mussels may grow about the plants so abundantly as to obscure their basal attachment to the rocks. Both the stems and the repeatedly subdivided branches of the seaweed are astir with life, but with life on so small a scale that the human eye can see its details only with the aid of a microscope.

Snails, some with brightly banded and deeply sculptured shells, crawl along the fronds, browsing on microscopic vegetable matter. Many of the basal stems of the weed are thickly encrusted with the bryozoan sea lace, Membranipora; from all its compartments the minute, be-tentacled heads of the resident creatures are thrust out. Another bryozoan of coarser growth, Flustrella, also forms mats investing the broken stems and stubble of the red weed, the substance of its own growth giving such a stem almost the thickness of a pencil. Rough hairs or bristles protrude from the mat, so that much foreign matter adheres to it. Like the sea laces, however, it is formed of hundreds of small, adjacent compartments. From one after another of these, as I watch through my microscope, a stout little creature cautiously emerges, then unfurls its crown of filmy tentacles as one would open an umbrella. Threadlike worms creep over the bryozoan, winding among the bristles like snakes through coarse stubble. A tiny, cyclopean crustacean, with one glittering ruby eye, runs ceaselessly and rather clumsily over the colony, apparently disturbing the inhabitants, for when

Mussels

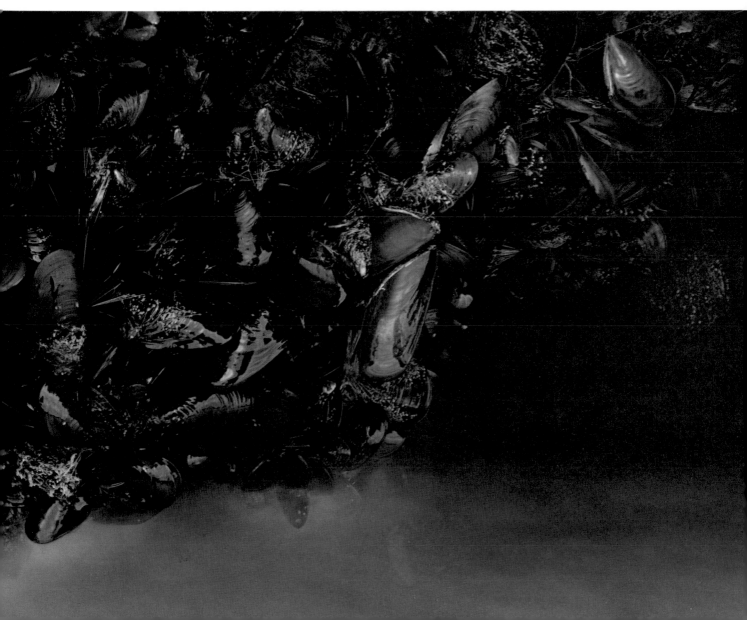

one of them feels the touch of the blundering crustacean it quickly folds its tentacles and withdraws into its compartment.

In the upper branches of this jungle formed by the red weed, there are many nests or tubes occupied by amphipod crustaceans known as Amphithoë. These small creatures have the appearance of wearing cream-colored jerseys brightly splotched with brownish red; in each goat-like face are set two conspicuous eyes and two pairs of hornlike antennae. The nests are as firmly and skillfully constructed as a bird's but are subject to far more continuous use, for these amphipods are weak swimmers and ordinarily seem loath to leave their nests. They lie in their snug little sacs, often with the heads and upper parts of their bodies protruding. The water currents that pass through their seaweed home bring them small plant fragments and thus solve the problem of subsistence.

For most of the year Amphithoë live singly, one to a nest. Early in the summer the males visit the females (who greatly outnumber them) and mating occurs within the nest. As the young develop the mother cares for them in a brood-pouch formed by the appendages of her abdomen. Often, while carrying her young, she emerges almost completely from her nest and vigorously fans currents of water through the pouch.

The eggs yield embryos, the embryos become larvae; but still the mother holds and cares for them until their small bodies have so developed that they are able to set forth into the seaweeds, to spin their own nests out of the fibers of plants and the silken threads mysteriously fashioned in their own bodies, and to feed and fend for themselves.

As her young become ready for independent life, the mother shows impatience to be rid of the swarm in her nest. Using claws and antennae, she pushes them to the rim and with shoves and nudges tries to expel them. The young cling with hooked and bristled claws to the walls and doorway of the familiar nursery. When finally thrust out they are likely to linger nearby; when the mother incautiously emerges, they leap to attach themselves to her body and so be drawn again into the security of their accustomed nest, until maternal impatience once more becomes strong.

Even the young just out of the brood-sac build their own nests and enlarge them as their growth requires. But the young seem to spend less time than the adults do inside their nests, and to creep about more freely over the weeds. It is common to see several tiny nests built close to the home of a large amphipod; perhaps the young like to stay close to the mother even after they have been ejected from her nest.

At low tide the water falls below the rockweeds and the mussels and enters a broad band clothed with the reddish-brown turf of the Irish moss. The time of its exposure to the atmosphere is so brief, the retreat of the sea so fleeting, that the moss retains a shining freshness, a wetness, and a sparkle that speak of its recent contact with the surf. Perhaps because we can visit this area only in that brief and magical hour of the tide's turning, perhaps because of the nearness of waves breaking on rocky rims, dissolving in foam and spray, and pouring seaward again to the accompaniment of many water sounds, we are reminded always that this low-tide area is of the sea and that we are trespassers.

Here, in this mossy turf, life exists in layers, one above another; life exists on other life, or within it, or under it, or above it. Because the moss is low-growing and branches profusely and intricately, it cushions the living things within it from the blows of the surf, and holds the wetness of the sea about them in these brief intervals of the low ebbing of the tide. After I have visited the shore and then at night have heard the surf trampling in over these moss-grown ledges with the heavy tread of the fall tides, I have wondered about the baby starfish, the urchins, the brittle stars, the tube-dwelling amphipods, the nudibranchs, and all the other small and delicate fauna of the moss; but I know that if there is security in their world it should be here, in this densest of intertidal jungles, over which the waves break harmlessly.

The moss forms so dense a covering that one cannot see what is beneath without intimate exploration. The abundance of life here, both in species and individuals, is on a scale that is hard to grasp. There is scarcely a stem of Irish moss that is not completely encased with one of the bryozoan sea mats—the white lacework of Membranipora or the glassy,

Rocks covered with Irish moss

brittle crust of Microporella. Such a crust consists of a mosaic of almost microscopic cells or compartments, arranged in regular rows and patterns, their surfaces finely sculptured. Each cell is the home of a minute, tentacled creature. By a conservative guess, several thousand such creatures live on a single stem of moss. On a square foot of rock surface there are probably several hundred such stems, providing living space for about a million of the bryozoans. On a stretch of Maine shore that the eye can take in at a glance, the population must run into the trillions for this single group of animals.

But there are further implications. If the population of the sea laces is so immense, that of the creatures they feed upon must be infinitely greater. A bryozoan colony acts as a highly efficient trap or filter to remove minute food animals from the sea water. One by one, the doors of the separate compartments open and from each a whorl of petal-like filaments is thrust out. In one moment the whole surface of the colony may be alive with crowns of tentacles swaying like flowers in a wind-swept field; the next instant, all may have snapped back into their protective cells and the colony is again a pavement of sculptured stone. But while the "flowers" sway over the stone field each spells death for many beings of the sea, as it draws in the minute spheres and ovals and crescents of the protozoans and the smallest algae, perhaps also some of the smallest of crustaceans and worms, or the larvae of mollusks and starfish, all of which are invisibly present in this mossy jungle, in numbers like the stars.

Larger animals are less numerous but still impressively abundant. Sea urchins, looking like large green cockleburs, often lie deep within the moss, their globular bodies anchored securely to the underlying rock by the adhesive discs of many tube feet. The ubiquitous common periwinkles, in some curious way unaffected by the conditions that confine most intertidal animals to certain zones, live above, within, and below the moss zone. Here their shells lie about over the surface of the weed at low tide; they hang heavily from its fronds, ready to drop at a touch.

And young starfish are here by the hundred, for these meadows of moss seem to be one of the chief nurseries for the starfish of northern shores. In the fall almost every other plant shelters quarter-inch and half-inch sizes. In these youthful starfish there are color patterns that become obliterated in maturity. The tube feet, the spines, and all the other curious epidermal outgrowths of these spiny-skinned creatures are large in proportion to the total size and have a clean perfection of form and structure.

On the rocky floor among the plant stems lie the infant stars. They are white insubstantial specks, in size and delicate beauty like snowflakes. There is an obvious newness about them, proclaiming that they have undergone their metamorphosis from the larval form to the adult shape only recently.

Perhaps it was on these very rocks that the swimming larvae, completing their period of life in the plankton, came to rest, attaching themselves firmly and becoming for a brief period sedentary animals. Then their bodies were like blown glass from which slender horns projected; the horns or lobes were covered with cilia for swimming and some of them bore suckers for use when the larvae should seek the firm underlying floor of the sea. During the short but critical period of attachment, the larval tissues were reorganized as completely as those of a pupal insect within a cocoon, the infant shape disappeared and in its place the five-rayed body of the adult was formed. Now as we find them, these new-made starfish use their tube feet competently, creeping over the rocks, righting their bodies if by mischance they are overturned, even, we may suppose, finding and devouring minute food animals in true starfish fashion.

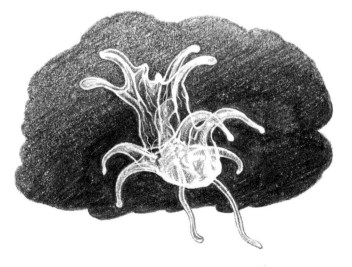

Swimming larva of starfish

The northern starfish lives in almost every low-tide pool or waits out the tidal interval in wet moss or in the dripping coolness of a rock overhang. On a very low tide, when the departure of the sea is brief, these stars strew their variously hued forms over the moss like so many blossoms—pink, blue, purple, peach, or beige. Here and there is a gray or orange starfish on which the spines stand out conspicuously in a pattern of white dots. Its arms are rounder and firmer than those of the northern star and the round stony plate on its upper surface is usually a bright orange instead of pale yellow as in the northern species. This starfish is common south of Cape Cod and only a few individuals stray farther north. Still a third species inhabits these low-tide rocks—the blood-red starfish, Henricia, whose kind not only lives at these margins of the sea but goes down to lightless sea bottoms near the edge of the continental shelf. It is always an inhabitant of cool waters and south of Cape Cod must go offshore to find the temperatures it requires. But its dispersal is not, as one might suppose, by the larval stages, for unlike most other starfish it produces no swimming young; instead, the mother holds the eggs and the young that develop from them in a pouch formed by her arms as she assumes a humped position. Thus she broods them until they have become fully developed little starfish.

The Jonah crabs use the resilient cushion of moss as a hiding place to wait for the return of the tide or the coming of darkness. I remember a moss-carpeted ledge standing out from a rock wall, jutting out over sea depths where laminaria rolled in the tide. The sea had only recently dropped below this ledge; its return was imminent and in fact was promised by every glassy swell that surged smoothly to its edge, then fell away. The moss was saturated, holding the water as faithfully as a sponge. Down within the deep pile of that carpet I caught a glimpse of a bright rosy color. At first I took it to be a growth of one of the encrusting corallines, but when I parted the fronds I was startled by abrupt movement as a large crab shifted its position and lapsed again into passive waiting. Only after search deep in the moss did I find several of the crabs, waiting out the brief interval of low tide and reasonably secure from detection by the gulls.

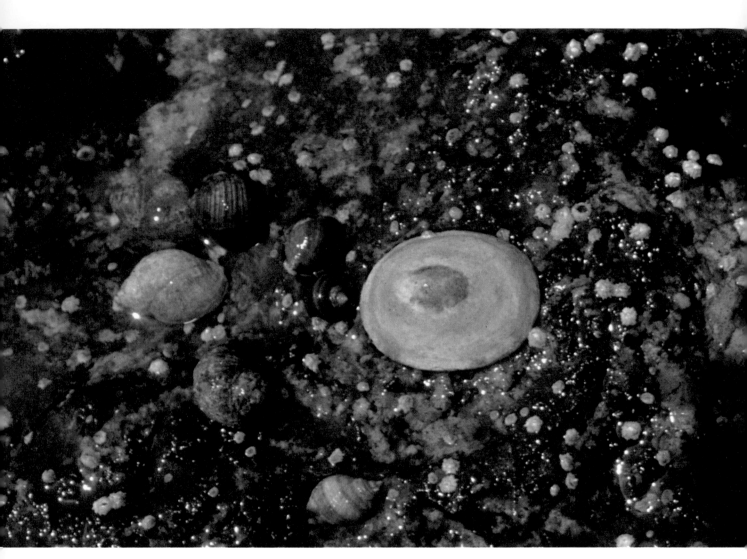

Limpet

The seeming passivity of these northern crabs must be related to their need to escape the gulls—probably their most persistent enemies. By day one always has to search for the crabs. If not hidden deeply within the seaweeds, they may be wedged in the farthest recess afforded by an overhanging rock, secure there, in dim coolness, gently waving their antennae and waiting for the return of the sea. In darkness, however, the big crabs possess the shore. One night when the tide was ebbing I went down to the low-tide world to return a large starfish I had taken on the morning tide. The starfish was at home at the lowest level of these tides of the August moon, and to that level it must be returned. I took a flashlight and made my way down over the slippery rockweeds. It was an eerie world; ledges curtained with weed and boulders that by day were familiar landmarks seemed to loom larger than I remembered and to have assumed unfamiliar shapes, every projecting mass thrown into bold relief by the shadows. Everywhere I looked, directly in the beam of my flashlight or obliquely in the half-illuminated gloom, crabs were scuttling about. Boldly and possessively they inhabited the weed-shrouded rocks. All the grotesqueness of their form accentuated, they seemed to have transformed this once familiar place into a goblin world.

In some places, the moss is attached, not to the underlying rock, but to the next lower layer of life, a community of horse mussels. These large mollusks inhabit heavy, bulging shells, the smaller ends of which bristle with coarse yellow hairs that grow as excrescences from the epidermis. The horse mussels themselves are the basis of a whole community of animals that would find life on these wave-swept rocks impossible except for the presence and activities of the mollusks. The mussels have bound their shells to the underlying rock by an almost inextricable tangle of golden-hued byssus threads. These are the product of glands in the long slender foot, the threads being "spun" from a curious milky secretion that solidifies on contact with sea water. The threads posses a texture that is a remarkable combination of toughness, strength, softness, and elasticity; extending out in all directions they enable the mussels to hold their position not only against the thrust of incoming waves but also against the drag of the backwash, which in a heavy surf is tremendous.

Over the years that the mussels have been growing here, particles of muddy debris have settled under their shells and around the anchor lines of the byssus threads. This has created still another area for life, a sort of understory inhabited by a variety of animals including worms, crustaceans, echinoderms, and numerous mollusks, as well as the baby mussels of an oncoming generation—these as yet so small and transparent that the forms of their infant bodies show through newly formed shells.

Certain animals almost invariably live among the horse mussels. Brittle stars insinuate their thin bodies among the threads and under the shells of the mussels, gliding with serpentine motions of the long slender arms. The scale worms always live here, too, and down in the lower layers of this strange community of animals starfish may live below the scale worms and brittle stars, and sea urchins below the starfish, and sea cucumbers below the urchins.

Of the echinoderms that live here, few are the largest individuals of their species. The blanket of horse mussels seems to be a shelter for young, growing animals, and indeed the full-grown starfish and urchins could hardly be accommodated there. In the waterless intervals of the low tide, the cucumbers draw themselves into little football-shaped ovals scarcely more than an inch long; returned to the water and fully relaxed, they extend their bodies to a length of five or six inches and unfurl a crown of tentacles. The cucumbers are detritus feeders, and explore the surrounding muddy debris with their soft tentacles, which periodically they pull back and draw across their mouths, as a child would lick his fingers.

In pockets deep in the moss under layers of mussels, a long, slender little fish of the blenny tribe, the rock eel, waits for the return of the tide, coiled in its water-filled refuge with several of its kind. Disturbed by an intruder, all thrash the water violently, squirming with eel-like undulations to escape.

Where the big mussels grow more sparsely, in the seaward suburbs of this mussel city, the moss carpet, too, becomes a little thinner; but still the underlying rock seldom is exposed. The green crumb-of-bread sponge, which at higher levels seeks the shelter of rock overhangs and tide pools,

Crumb-of-bread sponge;
lower right: *brittle star hunting*
food in sponge

here seems able to face the direct force of the sea and forms soft, thick mats of pale green, dotted with the cones and craters typical of this species. And here and there patches of another color show amid the thinning moss—dull rose or a gleaming, reddish brown of satin finish—an intimation of what lies at lower levels.

During much of the year the spring tides drop down into the band of Irish moss but go no lower, returning then toward the land. But in certain months, depending on the changing positions of sun and moon and earth, even the spring tides gain in amplitude, and their surge of water ebbs farther into the sea even as it rises higher against the land. Always, the autumn tides move strongly, and as the hunter's moon waxes and grows round, there come days and nights when the flood tides leap at the smooth rim of granite and send up their lace-edged wavelets to touch the roots of the bayberry; on their ebbs, with sun and moon combining to draw them back to the sea, they fall away from ledges not revealed since the April moon shone upon their dark shapes. Then they expose the sea's enameled floor—the rose of encrusting corallines, the green of sea urchins, the shining amber of the oarweeds.

At such a time of great tides I go down to that threshold of the sea world to which land creatures are admitted rarely in the cycle of the year. There I have known dark caves where tiny sea flowers bloom and masses of soft coral endure the transient withdrawal of the water. In these caves and in the wet gloom of deep crevices in the rocks I have found myself in the world of the sea anemones—creatures that spread a creamy-hued

91

Sea urchins and coralline algae

crown of tentacles above the shining brown columns of their bodies, like handsome chrysanthemums blooming in little pools held in depressions or on bottoms just below the tide line.

Where they are exposed by this extreme ebbing of the water, their appearance is so changed that they seem not meant for even this brief experience of land life. Wherever the contours of this uneven sea floor provide some shelter I have found their exposed colonies—dozens or scores of anemones crowded together, their translucent bodies touching, side against side. The anemones that cling to horizontal surfaces respond to the withdrawal of water by pulling all their issues down into a flattened, conical mass of firm consistency. The crown of feather-soft tentacles is retracted and tucked within, with no suggestion of the beauty that resides in an expanded anemone. Those that grow on vertical rocks hang down limply, extended into curious, hourglass shapes, all their tissues flaccid in the unaccustomed withdrawal of water. They do not lack the ability to contract, for when they are touched they promptly begin to shorten the column, drawing it up into more normal proportions. These anemones, deserted by the sea, are bizarre objects rather than things of beauty, and indeed bear only the most remote resemblance to the anemones blooming under water just offshore, all their tentacles expanded in the search for food. As small water creatures come in contact with the tentacles of these expanded anemones, they receive a deadly discharge. Each of the thousand or more tentacles bears thousands of coiled darts embedded in its substance, each with a minute spine protruding. The spine may act as a trigger to set off the explosion, or perhaps the very nearness of prey acts as a sort of chemical trigger, causing the dart to explode with great violence, impaling or entangling its victim and injecting a poison.

Like the anemones, the soft coral hangs its thimble-sized colonies on the under side of ledges. Limp and dripping at low tide, they suggest nothing of the life and beauty to which the returning water restores them. Then from all the myriad pores of the surface of the colony, the tentacles of little tubular animals appear and the polyps thrust themselves out into the tide, seizing each for itself the minute shrimps and copepods and multiformed larvae brought by the water.

The soft coral, or sea finger, secretes no limy cups as the distantly related stony, or reef, corals do, but forms colonies in which many animals live embedded in a tough matrix strengthened with spicules of lime. Minute though the spicules are, they become geologically important where, in tropical reefs, the soft corals, or Alcyonaria, mingle with the true corals. With the death and dissolution of the soft tissues, the hard spicules become minute building stones, entering into the composition of the reef. Alcyonarians grow in lush profusion and variety on the coral reefs and flats of the Indian Ocean, for these soft corals are predominantly creatures of the tropics. A few, however, venture into polar waters. One very large species, tall as a tall man and branched like a tree, lives on the fishing banks off Nova Scotia and New England. Most of the group live in deep waters; for the most part the intertidal rocks are inhospitable to them and only an occasional low-lying ledge, rarely and briefly exposed on the low spring tides, bears their colonies on dark and hidden surfaces.

In seams and crevices of rock, in little water-filled pools, or on rock walls briefly exposed by the tide's low ebbing, colonies of the pink-hearted hydroid Tubularia form gardens of beauty. Where the water still covers them the flowerlike animals sway gracefully at the ends of long stalks, their tentacles reaching out to capture small animals of the plankton. Perhaps it is where they are permanently submerged, however, that they reach their fullest development. I have seen them coating wharf pilings, floats, and submerged ropes and cables so thickly that not a trace of the substratum could be seen, their growth giving the illusion of thousands of blossoms, each as large as the tip of my little finger.

Below the last clumps of Irish moss, a new kind of sea bottom is exposed. The transition is abrupt. As though a line had been drawn, suddenly there is no more moss, and one steps from the yielding brown cushion onto a surface that seemingly is of stone. Except that the color is wrong, the effect is almost that of a volcanic slope—there is the same barren nakedness. Yet this is not rock that we see. The underlying rock is coated on every surface, vertical or horizontal, exposed or hidden, with a crust of coralline algae, so that it wears a rich old-rose color. So intimate is

Skeleton shrimp on hydroid Tubularia

the union that the plant seems part of the rock. Here the periwinkles wear little patches of pink on their shells, all the rock caverns and fissures are lined with the same color, and the rock bottom that slants away into green water carries down the rose hue as far as the eye can follow.

The coralline algae are plants of unusual fascination. They belong to the group of red seaweeds, most of which live in the deeper coastal waters, for the chemical nature of their pigments usually requires the protection of a screen of water between their tissues and the sun. The corallines, however, are extraordinary in their ability to withstand direct sunlight. They are able to incorporate carbonate of lime into their tissues so that they have become hardened. Most species form encrusting patches on rocks, shells, and other firm surfaces. The crust may be thin and smooth, suggesting a coat of enamel paint; or it may be thick and roughened by small nodules and protuberances. In the tropics the corallines often enter importantly into the composition of the coral reefs, helping to cement the branching structures built by the coral animals into a solid reef. Here and there in the East Indies they cover the tidal flats as far as eye can see with their delicately hued crusts, and many of the "coral reefs" of the Indian Ocean contain no coral but are built largely of these plants. About the coasts of Spitsbergen, where under the dimly lit waters of the north the great forests of the brown algae grow, there are also vast calcareous banks, stretching mile after mile, formed by the coralline algae. Being able to live

95

not only in tropical warmth but where water temperatures seldom rise above the freezing point, these plants flourish all the way from Arctic to Antarctic seas.

Where these same corallines paint a rose-colored band on the rocks of the Maine coast, as though to mark the low-water line of the lowest spring tides, visible animal life is scarce. But although little else lives openly in this zone, thousands of sea urchins do. Instead of hiding in crevices or under rocks as they do at the higher levels, they live fully exposed on the flat or gently shelving rock faces. Groups of a score or half a hundred individuals lie together on the coralline-coated rocks, forming patches of pure green on the rose background. I have seen such herds of urchins lying on rocks that were being washed by a heavy surf, but apparently all the little anchors formed by their tube feet held securely. Though the waves broke heavily and poured back in a turbulent rush of waters, there the urchins remained undisturbed. Perhaps the strong tendency to hide and to wedge themselves into crevices and under boulders, as displayed by urchins in tide pools or up in the rockweed zone, is not so much an attempt to avoid the power of the surf as a means of escaping the eager eyes of the gulls, who hunt them relentlessly on every low tide. This coralline zone where the urchins live so openly is covered almost constantly with a protective layer of water; probably not more than a dozen daytime tides in the entire year fall to this level. At all other times, the depth of water over the urchins prevents the gulls from reaching them, for although a gull can make shallow plunges under water, it cannot dive as a tern does, and probably cannot reach a bottom deeper than the length of its own body.

The lives of many of these creatures of the low-tide rocks are bound together by interlacing ties, in the relation of predator to prey, or in the relation of species that compete for space or food. Over all these the sea itself exercises a directing and regulating force.

The urchins seek sanctuary from the gulls at this low level of the spring tides, but in themselves stand in the relation of dangerous predators to other animals. Where they advance into the Irish moss zone, hiding in

Urchin test

deep crevices and sheltering under rock overhangs, they devour numbers of periwinkles, and even attack barnacles and mussels. The number of urchins at any particular level of shore has a strong regulating effect on the populations of their prey. The starfish and a voracious snail, the common whelk, like the urchins, have their centers of population in deep water offshore and make predatory excursions of varying duration into the intertidal zone.

The position of the prey animals—the mussels, barnacles, and periwinkles—on sheltered shores has become difficult. They are hardy and adaptable, able to live at any level of the tide. Yet on such shores the rockweeds have crowded them out of the upper two-thirds of the shore, except for scattered individuals. At and just below the low-tide line are the hungry predators, so all that remains for these animals is the level near the low-water line of the neap tides. On protected coasts it is here that the barnacles and mussels assemble in their millions to spread their cover of white and blue over the rocks, and the legions of the common periwinkle gather.

But the sea, with its tempering and modifying effect, can alter the pattern. Whelks, starfish, and urchins are creatures of cold water. Where the offshore waters are cold and deep and the tidal flow is drawn from these icy reservoirs, the predators can range up into the intertidal zone, decimating the numbers of their prey. But when there is a layer of warm surface water the predators are confined to the cold deep levels. As they retreat seaward, the legions of their prey follow down in their wake, descending as far as they may into the world of the low spring tides.

Tide pools contain mysterious worlds within their depths, where all the beauty of the sea is subtly suggested and portrayed in miniature. Some of the pools occupy deep crevices or fissures; at their seaward ends these crevices disappear under water, but toward the land they run back slantingly into the cliffs and their walls rise higher, casting deep shadows over the water within them. Other pools are contained in rocky basins with a high rim on the seaward side to hold back the water when the last

of the ebb drains away. Seaweeds line their walls. Sponges, hydroids, anemones, sea slugs, mussels, and starfish live in water that is calm for hours at a time, while just beyond the protecting rim the surf may be pounding.

The pools have many moods. At night they hold the stars and reflect the light of the Milky Way as it flows across the sky above them. Other, living stars come in from the sea: the shining emeralds of tiny phosphorescent diatoms—the glowing eyes of small fishes that swim at the surface of the dark water, their bodies slender as matchsticks, moving almost upright with little snouts uplifted—the elusive moonbeam flashes of comb jellies that have come in with a rising tide. Fishes and comb jellies hunt the black recesses of the rock basins, but like the tides they come and go, having no part in the permanent life of the pools.

By day there are other moods. Some of the most beautiful pools lie high on the shore. Their beauty is the beauty of simple elements—color and form and reflection. I know one that is only a few inches deep, yet it holds all the depth of the sky within it, capturing and confining the reflected blue of far distances. The pool is outlined by a band of bright green, a growth of one of the seaweeds called Enteromorpha. The fronds of the weed are shaped like simple tubes or straws. On the land side a wall of gray rock rises above the surface to the height of a man, and reflected, descends its own depth into the water. Beyond and below the reflected cliff are those far reaches of the sky. When the light and one's mood are right, one can look down into the blue so far that one would hesitate to set foot in so bottomless a pool. Clouds drift across it and wind ripples scud over its surface, but little else moves there, and the pool belongs to the rock and the plants and the sky.

In another high pool nearby, the green tube-weed rises from all of the floor. By some magic the pool transcends its realties of rock and water and plants, and out of these elements creates the illusion of another world. Looking into the pool, one sees no water but instead a pleasant landscape of hills and valleys with scattered forests. Yet the illusion is not so much that of an actual landscape as of a painting of one; like the strokes of a

skillful artist's brush, the individual fronds of the algae do not literally portray trees, they merely suggest them. But the artistry of the pool, as of the painter, creates the image and the impression.

Little or no animal life is visible in any of these high pools—perhaps a few periwinkles and a scattering of little amber isopods. Conditions are difficult in all pools high on the shore because of the prolonged absence of the sea. The temperature of the water may rise many degrees, reflecting the heat of the day. The water freshens under heavy rains or becomes more salty under a hot sun. It varies between acid and alkaline in a short time through the chemical activity of the plants. Lower on the shore the pools provide far more stable conditions, and both plants and animals are able to live at higher levels than they could on open rock. The tide pools, then, have the effect of moving the life zones a little higher on the shore. Yet they, too, are affected by the duration of the sea's absence, and the inhabitants of a high pool are very different from those of a low-level pool that is separated from the sea only at long intervals and then briefly.

The highest of the pools scarcely belong to the sea at all; they hold the rains and receive only an occasional influx of sea water from storm surf or very high tides. But the gulls fly up from their hunting at the sea's edge, bringing a sea urchin or a crab or a mussel to drop on the rocks, in this way shattering the hard shelly covering and exposing the soft parts within. Bits of urchin tests or crab claws or mussel shells find their way into the pools, and as they disintegrate their limy substance enters into the chemistry of the water, which then becomes alkaline. A little one-celled plant called Sphaerella finds this a favorable climate for growth—a minute, globular bit of life almost invisible as an individual, but in its millions turning the waters of these high pools red as blood. Apparently the alkalinity is a necessary condition; other pools, outwardly similar except for the chance circumstance that they contain no shells, have none of the tiny crimson balls.

Even the smallest pools, filling depressions no larger than a teacup, have some life. Often it is a thin patch of scores of the little seashore insect, *Anurida maritima*—"the wingless one who goes to sea." These

High-level tide pool, surrounded by bits of crab shells left by gulls

High pool stained red by Sphaerella

small insects run on the surface film when the water is undisturbed, crossing easily from one shore of a pool to another. Even the slightest rippling causes them to drift helplessly, however, so that scores or hundreds of them come together by chance, becoming conspicuous only as they form thin, leaflike patches on the water. A single Anurida is small as a gnat. Under a lens, it seems to be clothed in blue-gray velvet through which many bristles or hairs protrude. The bristles hold a film of air about the body of the insect when it enters the water, and so it need not return to the upper shore when the tide rises. Wrapped in its glistening air blanket, dry and provided with air for breathing, it waits in cracks and crevices until the tide ebbs again. Then it emerges to roam over the rocks, searching for the bodies of fish and crabs and the dead mollusks and barnacles that provide its food, for it is one of the scavengers that play a part in the economy of the sea, keeping the organic materials in circulation.

And often I find the pools of the upper third of the shore lined with a brown velvety coating. My fingers, exploring, are able to peel it off the rocks in thin smooth-surfaced sheets like parchment. It is one of the brown seaweeds called Ralfsia; it appears on the rocks in small, lichen-like growths or, as here, spreading its thin crust over extensive areas. Wherever it grows its presence changes the nature of a pool, for it provides the shelter that many small creatures seek so urgently. Those small enough to

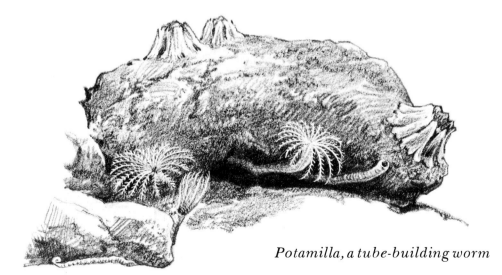

Potamilla, a tube-building worm

creep in under it—to inhabit the dark pockets of space between the encrusting weed and the rock—have found security against being washed away by the surf. Looking at these pools with their velvet lining, one would say there is little life here—only a sprinkling of periwinkles browsing, their shells rocking gently as they scrape at the surface of the brown crust, or perhaps a few barnacles with their cones protruding through the sheet of plant tissue, opening their doors to sweep the water for food. But whenever I have brought a sample of this brown seaweed to my microscope, I have found it teeming with life. Always there have been many cylindrical tubes, needle-fine, built of a muddy substance. The architect of each is a small worm whose body is formed of a series of eleven infinitely small rings or segments, like eleven counters in a game of checkers, piled one above another. From its head arises a structure that makes this otherwise drab worm beautiful—a fanlike crown or plume composed of the finest feathery filaments. The filaments absorb oxygen and also serve to ensnare small food organisms when thrust out of the tube. And always, among this microfauna of the Ralfsia crust, there have been little fork-tailed crustaceans with glittering eyes the color of rubies. Other crustaceans called ostracods are enclosed in flattened, peach-colored shells fashioned of two parts, like a box with its lid; from the shell long appendages may be thrust out to row the creatures through the water. But most numerous of all are the minute worms hurrying across the crust— segmented bristle worms of many species and smooth-bodied, serpent-like ribbon worms or nemerteans, their appearance and rapid movements betraying their predatory errands.

A pool need not be large to hold beauty within pellucid depths. I remember one that occupied the shallowest of depressions; as I lay outstretched on the rocks beside it I could easily touch its far shore. This miniature pool was about midway between the tide lines, and for all I could see it was inhabited by only two kinds of life. Its floor was paved with mussels. Their shells were a soft color, the misty blue of distant mountain ranges, and their presence lent an illusion of depth. The water in which they lived was so clear as to be invisible to my eyes; I could detect the interface between air and water only by the sense of coldness on my fingertips. The crystal water was filled with sunshine—an infusion and distillation of light that reached down and surrounded each of these small but resplendent shellfish with its glowing radiance.

The mussels provided a place of attachment for the only other visible life of the pool. Fine as the finest threads, the basal stems of colonies of hydroids traced their almost invisible lines across the mussel shells. The hydroids belonged to the group called Sertularia, in which each individual of the colony and all the supporting and connecting branches are enclosed within transparent sheaths, like a tree in winter wearing a sheath of ice. From the basal stems erect branches arose, each branch the bearer of a double row of crystal cups within which the tiny beings of the colony dwelt. The whole was the very embodiment of beauty and fragility, and as I lay beside the pool and my lens brought the hydroids into clearer view they seemed to me to look like nothing so much as the finest cut glass—perhaps the individual segments of an intricately wrought chandelier. Each animal in its protective cup was something like a very small sea anemone—a little tubular being surmounted by a crown of tentacles. The central cavity of each communicated with a cavity that ran the length of the branch that bore it, and this in turn with the cavities of larger branches and with those of the main stem, so that the feeding activities of each animal contributed to the nourishment of the whole colony.

On what, I wondered, were these Sertularians feeding? From their very abundance I knew that whatever creatures served them as food must be infinitely more numerous than the carnivorous hydroids themselves.

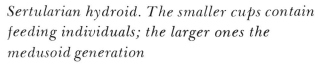
Sertularian hydroid. The smaller cups contain feeding individuals; the larger ones the medusoid generation

Yet I could see nothing. Obviously their food would be minute, for each of the feeders was of threadlike diameter and its tentacles were like the finest gossamer. Somewhere in the crystal clarity of the pool my eye—or so it seemed—could detect a fine mist of infinitely small particles, like dust motes in a ray of sunshine. Then as I looked more closely the motes had disappeared and there seemed to be once more only that perfect clarity, and the sense that there had been an optical illusion. Yet I knew it was only the human imperfection of my vision that prevented me from seeing those microscopic hordes that were the prey of the groping, searching tentacles I could barely see. Even more than the visible life, that which was unseen came to dominate my thoughts, and finally the invisible throng seemed to me the most powerful beings in the pool. Both the hydroids and the mussels were utterly dependent on this invisible flotsam of the tide streams, the mussels as passive strainers of the plant plankton, the hydroids as active predators seizing and ensnaring the minute water fleas and copepods and worms. But should the plankton become less abundant, should the incoming tide streams somehow become drained of this life, then the pool would become a pool of death, both for the mussels in their shells blue as mountains and for the crystal colonies of the hydroids.

Some of the most beautiful pools of the shore are not exposed to the view of the casual passer-by. They must be searched for—perhaps in low-lying basins hidden by great rocks that seem to be heaped in disorder and confusion, perhaps in darkened recesses under a projecting ledge, perhaps behind a thick curtain of concealing weeds.

I know such a hidden pool. It lies in a sea cave, at low tide filling perhaps the lower third of its chamber. As the flooding tide returns the pool grows, swelling in volume until all the cave is water-filled and the cave and the rocks that form and contain it are drowned beneath the fullness of the tide. When the tide is low, however, the cave may be approached from the landward side. Massive rocks form its floor and walls and roof. They are penetrated by only a few openings—two near the floor on the sea side and one high on the landward wall. Here one may lie on the rocky threshold and peer through the low entrance into the cave and down into its pool. The cave is not really dark; indeed on a bright day it glows with a cool green light. The source of this soft radiance is the sunlight that enters through the openings low on the floor of the pool, but only after its entrance into the pool does the light itself become transformed, invested with a living color of purest, palest green that is borrowed from the covering of sponge on the floor of the cave.

Through the same openings that admit the light, fish come in from the sea, explore the green hall, and depart again into the vaster waters beyond. Through those low portals the tides ebb and flow. Invisibly, they bring in minerals—the raw materials for the living chemistry of the plants and animals of the cave. They bring, invisibly again, the larvae of many sea creatures—drifting, drifting in their search for a resting place. Some may remain and settle here; others will go out on the next tide.

Looking down into the small world confined within the walls of the cave, one feels the rhythms of the greater sea world beyond. The waters of the pool are never still. Their level changes not only gradually with the rise and fall of the tide, but also abruptly with the pulse of the surf. As the backwash of a wave draws it seaward, the water falls away rapidly; then with a sudden reversal the inrushing water foams and surges upward almost to one's face.

On the outward movement one can look down and see the floor, its details revealed more clearly in the shallowing water. The green crumb-of-bread sponge covers much of the bottom of the pool, forming a thick-piled carpet built of tough little feltlike fibers laced together with glassy, double-pointed needles of silica—the spicules or skeletal supports of the sponge. The green color of the carpet is the pure color of chlorophyll, this plant pigment being confined within the cells of an alga that are scattered through the tissues of the animal host. The sponge clings closely to the rock, by the very smoothness and flatness of its growth testifying to the streamlining force of heavy surf. In quiet waters the same species sends up many projecting cones; here these would give the turbulent waters a surface to grip and tear.

Interrupting the green carpet are patches of other colors, one a deep,

110

Starfish on wall near entrance to a sea cave

mustard yellow, probably a growth of the sulphur sponge. In the fleeting
moment when most of the water has drained away, one has glimpses of a
rich orchid color in the deepest part of the cave—the color of the
encrusting coralline algae.

Sponges and corallines together form a background for the larger
tide-pool animals. In the quiet of ebb tide there is little or no visible
movement even among the predatory starfish that cling to the walls like

ornamental fixtures painted orange or rose or purple. A group of large anemones lives on the wall of the cave, their apricot color vivid against the green sponge. Today all the anemones may be attached on the north wall of the pool, seemingly immobile and immovable; on the next spring tides when I visit the pool again some of them may have shifted over to the west wall and there taken up their station, again seemingly immovable.

There is abundant promise that the anemone colony is a thriving one and will be maintained. On the walls and ceiling of the cave are scores of baby anemones—little glistening mounds of soft tissue, a pale, translucent brown. But the real nursery of the colony seems to be in a sort of antechamber opening into the central cave. There a roughly cylindrical space no more than a foot across is enclosed by high perpendicular rock walls to which hundreds of baby anemones cling.

On the roof of the cave is written a starkly simple statement of the force of the surf. Waves entering a confined space always concentrate all their tremendous force for a driving, upward leap: in this manner the roofs of caves are gradually battered away. The open portal in which I lie saves the ceiling of this cave from receiving the full force of such upward-leaping waves; nevertheless, the creatures that live there are exclusively a heavy-surf fauna. It is a simple black and white mosaic—the black of mussel shells, on which the white cones of barnacles are growing. For some reason the barnacles, skilled colonizers of surf-swept rocks though they be, seem to have been unable to get a foothold directly on the roof of the cave. Yet the mussels have done so. I do not know how this happened but I can guess. I can imagine the young mussels creeping in over the damp rock while the tide is out, spinning their silk threads that bind them securely, anchoring them against the returning waters. And then in time, perhaps, the growing colony of mussels gave the infant barnacles a foothold more tenable than the smooth rock, so that they were able to cement themselves to the mussel shells. However it came about, that is the way we find them now.

As I lie and look into the pool there are moments of relative quiet, in the intervals when one wave has receded and the next has not yet entered. Then I can hear the small sounds: the sound of water dripping from the

Limpet, barnacles, and corallines on tide pool floor

mussels on the ceiling or of water dripping from seaweeds that line the walls—small, silver splashes losing themselves in the vastness of the pool and in the confused, murmurous whisperings that emanate from the pool itself—the pool that is never quite still.

Then as my fingers explore among the dark red thongs of the dulse and push away the fronds of the Irish moss that cover the walls beneath me, I begin to find creatures of such extreme delicacy that I wonder how they can exist in this cave when the brute force of storm surf is unleashed within its confined space.

Adhering to the rock walls are thin crusts of one of the bryozoans, a form in which hundreds of minute, flask-shaped cells of a brittle structure, fragile as glass, lie one against another in regular rows to form a continuous crust. The color is a pale apricot; the whole seems an ephemeral creation that would crumble away at a touch, as hoarfrost before the sun.

A tiny spiderlike creature with long and slender legs runs about over the crust. For some reason that may have to do with its food, it is the same apricot color as the bryozoan carpet beneath it; the sea spider, too, seems the embodiment of fragility.

Another bryozoan of coarser, upright growth, Flustrella, sends up little club-shaped projections from a basal mat. Again, the lime-impregnated clubs seem brittle and glassy. Over and among them, innumerable little roundworms crawl with serpentine motion, slender as threads. Baby mussels creep in their tentative exploration of a world so new to them they have not yet found a place to anchor themselves by slender silken lines.

Exploring with my lens, I find many very small snails in the fronds of seaweed. One of them has obviously not been long in the world, for its pure white shell has formed only the first turn of the spiral that will turn many times upon itself in growth from infancy to maturity. Another, no larger, is nevertheless older. Its shining amber shell is coiled like a French horn and, as I watch, the tiny creature within thrusts out a bovine head and seems to be regarding its surroundings with two black eyes, small as the smallest pinpoints.

But seemingly most fragile of all are the little calcareous sponges that here and there exist among the seaweeds. They form masses of minute,

114

Starfish

upthrust tubes of vase-like form, none more than half an inch high. The wall of each is a mesh of fine threads—a web of starched lace made to fairy scale.

I could have crushed any of these fragile structures between my fingers—yet somehow they find it possible to exist here, amid the surging thunder of the surf that must fill this cave as the sea comes in. Perhaps the seaweeds are the key to the mystery, their resilient fronds a sufficient cushion for all the minute and delicate beings they contain.

But it is the sponges that give to the cave and its pool their special quality—the sense of a continuing flow of time. For each day that I visit the pool on the lowest tides of the summer they seem unchanged—the same in July, the same in August, the same in September. And they are the same this year as last, and presumably as they will be a hundred or a thousand summers hence.

Simple in structure, little different from the first sponges that spread their mats on ancient rocks and drew their food from a primordial sea, the sponges bridge the eons of time. The green sponge that carpets the floor of this cave grew in other pools before this shore was formed; it was old when the first creatures came out of the sea in those ancient eras of the Paleozoic, 300 million years ago; it existed even in the dim past before the first fossil record, for the hard little spicules—all that remains when the living tissue is gone—are found in the first fossil-bearing rocks, those of the Cambrian period.

So, in the hidden chamber of that pool, time echoes down the long ages to a present that is but a moment.

As I watched, a fish swam in, a shadow in the green light, entering the pool by one of the openings low on its seaward wall. Compared with the ancient sponges, the fish was almost a symbol of modernity, its fishlike ancestry traceable only half as far into the past. And I, in whose eyes the images of the two were beheld as though they were contemporaries, was a mere newcomer whose ancestors had inhabited the earth so briefly that my presence was almost anachronistic.

As I lay at the threshold of the cave thinking those thoughts, the surge of waters rose and flooded across the rock on which I rested. The tide was rising.

INDEX